D0312956

孫子

Sun Tzu FOR EXECUTION

HOW TO USE THE ART OF WAR TO GET RESULTS

RETIRÉ DE LA COLLECTION UNIVERSELLE
Bibliothèque et Archives nationales du Québec

STEVEN W. MICHAELSON
COAUTHOR OF *SUN TZU FOR SUCCESS*

AVON, MASSACHUSETTS

Copyright © 2007, Steven W. Michaelson.
All rights reserved.
This book, or parts thereof, may not be reproduced in any form without permission from the publisher; exceptions are made for brief excerpts used in published reviews.

Published by Adams Business,
an imprint of Adams Media, an F+W Publications Company
57 Littlefield Street, Avon, MA 02322. U.S.A.
www.adamsmedia.com

ISBN 10: 1-59869-052-3
ISBN 13: 978-1-59869-052-1

Printed in the United States of America.

J I H G F E D C B

Library of Congress Cataloging-in-Publication Data
is available from publisher.

This publication is designed to provide accurate and authoritative information with regard to the subject matter covered. It is sold with the understanding that the publisher is not engaged in rendering legal, accounting, or other professional advice. If legal advice or other expert assistance is required, the services of a competent professional person should be sought.
—From a *Declaration of Principles* jointly adopted by a Committee of the American Bar Association and a Committee of Publishers and Associations

Many of the designations used by manufacturers and sellers to distinguish their product are claimed as trademarks. Where those designations appear in this book and Adams Media was aware of a trademark claim, the designations have been printed with initial capital letters.

This book is available at quantity discounts for bulk purchases.
For information, please call 1-800-289-0963.

To
Sue, K.T., and Danny

Contents

Contents

Introduction

SUN TZU IS VIEWED as the ultimate strategist of Eastern thought. However, he also knew how to *execute* strategy. One famous story about Sun Tzu is the *Story of the Concubines*. Essentially, Sun Tzu was challenged by the king to demonstrate his theories about managing soldiers. The king gave Sun Tzu his 100 concubines and challenged him to apply his theories and practices of training soldiers to managing the concubines. The concubines, as you might expect, were untrained in the matters of marching and drills and performed poorly. Sun Tzu decided to try a more decisive tactic—he beheaded the king's favorite concubine. The rest of the concubines, realizing that Sun Tzu was, indeed, serious about having his directives carried out, immediately began to execute his orders exceedingly well.

The story is an extreme one—but not without lessons. It illustrates how good execution is the foundation for good strategy. For Sun Tzu an army that followed its instructions was a prerequisite for an army that accomplished its objectives.

In any industry, if you ask your customers what is important, they will inherently describe the basics. If you are a retailer? It doesn't matter what goods you are selling or how upscale the store is, customers want the basics of good service—fast checkout lines and clean floors. What if you are a manufacturer? Customers want the product to work. What if you are in a service industry? Customers want you to be on time, to answer their questions, or to perform well at some other basic service metric.

Great execution gives a company strategic options.

Lexus is made by Toyota. Toyotas have an excellent reputation for reliability. That executional strength helped "birth" Lexus—customers immediately trusted that a Lexus would likely be as reliable as all of the other Toyota vehicles. The great in-store execution of Whole Foods let them change their strategy over the years from being known primarily as an organic food store to now doing a huge business in carry-out prepared foods. Their executional strength let them develop their strategy and broaden their appeal.

You can't put good strategy on top of bad execution and expect the strategy to work. Creating better execution is a lot about working with people. Few of us can have significant accomplishments without working through people. Those people might be our customers, people who work for us, our peers, or people in other parts of the organization.

Thus the organization of this book is built largely around the people-management principles. This isn't about "soft skills"—though those are important. It's about basic principles that allow you to accomplish significant objectives.

I have managed in big organizations (Procter & Gamble, and Sara Lee Corporation) and in fast-growing smaller companies (as president of Internet retailer FreshDirect). I have held high-level management positions in highly successful organizations (Wegmans, a perennial *Fortune* top company to work for in America, and one of the most successful and differentiated food retailers in the country) and in turnaround situations (Borden, at one time the world's largest dairy company but since broken up and sold in pieces). The principles of managing execution, people, and change do not vary from situation to situation.

Thus, this book is organized in five sections:

Section 1: Simplicity
Section 2: Create Alignment
Section 3: People Always
Section 4: Flexibility
Section 5: Have Towering Strengths

It is organized that way primarily because Sun Tzu's thoughts, when organized around execution, naturally organize in this way. Additionally, this structure matches my practical experience leading execution in a variety of organizations.

Simplicity is the first principle. If execution is more complicated than it needs to be, it won't be as successful. Maybe it won't be well understood. Maybe it won't be readily possible to do well. In business, keeping things simple is about keeping them easy. *Easy* is a good word. I can regularly perform, and perform well, things that are easy. So can others.

There are lots of ways to make things easy. I can take lessons. I can practice. I can buy or build something that makes a task easier. Easy is not the same as laziness. Work hard, but make things easy when you can.

As you simplify an operation, alignment of people and processes becomes easier. It also enables you to build greater executional strength. Alignment is about people and the tasks they need to accomplish. Thus, people-management skills are critical to ensuring execution. Also, all of these principles will achieve greater success if there is flexibility built into them.

Building execution is less a sequential process than an iterative process.

Build your execution toward some towering strengths. Strengths that are difficult for your competitors to copy. Building execution isn't free. It may take money—at a minimum it will take time. So build execution toward the right long-term end. This will provide a strong ultimate payout to your efforts.

This book's thoughts on execution are based on Sun Tzu's writings. Often referred to as one of the first management texts, Sun Tzu's writings remain timeless and versatile. Guiding principles frequently are. I hope you find insights from his thoughts as well.

—*Steve Michaelson*

Preface to Book 1

About Sun Tzu and His Book

SUN TZU LIVED in about 500 B.C. He is said to have written *The Art of War* on bamboo strips.

Sun Tzu is generally believed to have been a general in his day, though some believe that he was a civilian strategist. Still others deny his existence, claiming *The Art of War* was written by someone else.

Sun Tzu's writings have been highly influential, and over the years have developed into the foundation of Eastern military thought, in the same way Carl Von Clausewitz's writings are the foundation of much Western military thought. The *Los Angeles Herald Examiner* says, "Some of Mao Tse Tung's most eloquent thoughts are merely rehashes of Sun Tzu and his interpreters." Japan's Admiral Yamamoto, who planned the attack on Pearl Harbor, is said to have studied Sun Tzu.

In the business world, *The Art of War* has been called "the modern manager's bible" in an article in *Inc.* magazine. It has also figured prominently in the movie *Wall Street* and in one of the more recent James Bond 007 movies.

This translation of *The Art of War* is divided into thirteen chapters like the original work. The original thirteen chapters are divided into five sections here, to add a business relevance and context to Sun Tzu's writings.

BOOK ONE

Complete Translation of *The Art of War*

Chapter 1

Laying Plans

Thoroughly Assess Conditions

WAR IS A MATTER of vital importance to the state; a matter of life and death, the road either to survival or to ruin. Hence, it is imperative that it be thoroughly studied.

Therefore, to make assessment of the outcome of a war, one must compare the various conditions of the antagonistic sides in terms of the five constant factors:

1. Moral influence
2. Weather
3. Terrain
4. Commander
5. Doctrine

These five constant factors should be familiar to every general. He who masters them wins; he who does not is defeated.

Compare the Seven Attributes

Therefore, to forecast the outcome of a war, the attributes of the antagonistic sides should be analyzed by making the following seven comparisons:

1. Which sovereign possesses greater moral influence?
2. Which commander is more capable?
3. Which side holds more favorable conditions in weather and terrain?
4. On which side are decrees better implemented?
5. Which side is superior in arms?
6. On which side are officers and men better trained?
7. Which side is stricter and more impartial in meting out rewards and punishments?

By means of these seven elements, I can forecast victory or defeat.

If the sovereign heeds these stratagems of mine and acts upon them, he will surely win the war, and I shall, therefore, stay with him. If the sovereign neither heeds nor acts upon them, he will certainly suffer defeat, and I shall leave.

Look for Strategic Turns

Having paid attention to the advantages of my stratagems, the commander must create a helpful situation over and beyond the ordinary rules. By "situation," I mean he should act expediently in accordance with what is advantageous in the field and so meet any exigency.

All warfare is based on deception. Therefore, when able to attack, we must pretend to be unable; when employing our forces, we must seem inactive; when we are near, we must make the enemy believe we are far away; when far away, we must make him believe we are near.

Offer a bait to allure the enemy when he covets small advantages. Strike the enemy when he is in disorder. If he is well prepared with substantial strength, take double precautions against him. If he is powerful in action, evade him. If he is angry, seek to discourage him. If he appears humble, make him arrogant. If his forces have taken a good rest, wear them down. If his forces are united, divide them.

Launch the attack where he is unprepared; take action when it is unexpected.

These are the keys to victory for a strategist. However, it is impossible to formulate them in detail beforehand.

Now, the commander who gets many scores during the calculations in the temple before the war will have more likelihood of winning. The commander who gets few scores during the calculations in the temple before the war will have less chance of success. With many scores, one can win; with few scores, one cannot. How much less chance of victory has one who gets no scores at all! By examining the situation through these aspects, I can foresee who is likely to win or lose.

Chapter 2

Waging War

Marshal Adequate Resources

GENERALLY, OPERATIONS OF WAR involve 1,000 swift chariots, 1,000 heavy chariots, and 100,000 mailed troops, with the transportation of provisions for them over a thousand *li*. Thus, the expenditure at home and in the field, the stipends for the entertainment of state guests and diplomatic envoys, the cost of materials such as glue and lacquer, and the expense for care and maintenance of chariots and armor will amount to 1,000 pieces of gold a day. An army of 100,000 men can be raised only when this money is in hand.

Make Time Your Ally

In directing such an enormous army, a speedy victory is the main object.

If the war is long delayed, the men's weapons will be blunted and their ardor will be dampened. If the army attacks cities, their strength will be exhausted. Again, if the army engages in protracted campaigns, the resources of the state will not suffice. Now, when your weapons are blunted, your ardor dampened, your strength exhausted, and your treasure spent, neighboring rulers will take advantage of your distress to act. In this case, no man, however wise, is able to avert the disastrous consequences that ensue.

Thus, while we have heard of stupid haste in war, we have not yet seen a clever operation that was prolonged. There has never been a case in which a prolonged war has benefited a country. Therefore, only those who understand the dangers inherent in employing troops know how to conduct war in the most profitable way.

Everyone Must Profit from Victories

Those adept in employing troops do not require a second levy of conscripts or more than two provisionings. They carry military supplies from the homeland and make up for their provisions, relying on the enemy. Thus, the army will be always plentifully provided.

When a country is impoverished by military operations, it is because an army far from its homeland needs a distant transportation. Being forced to carry supplies for great distances renders the people destitute. On the other hand, the local price of commodities normally rises high in the area near the military camps. The rising prices cause financial resources to be drained away. When the resources are exhausted, the peasantry will be afflicted with urgent exactions. With this depletion of strength and exhaustion of wealth, every household in the homeland is left empty. Seven-tenths of the people's income is dissipated, and six-tenths of the government's revenue is paid for broken-down chariots, worn-out horses, armor and helmets, arrows and crossbows, halberds and bucklers, spears and body shields, draught oxen and heavy wagons.

Hence, a wise general is sure of getting provisions from the enemy countries. One *zhong* of grains obtained from the local area is equal to twenty *zhong* shipped from the home country;

one *dan* of fodder in the conquered area is equal to twenty *dan* from the domestic store.

Now, in order to kill the enemy, our men must be roused to anger; to gain the enemy's property, our men must be rewarded with war trophies. Accordingly, in chariot battle, when more than ten chariots have been captured, those who took the enemy chariot first should be rewarded. Then, the enemy's flags and banners should be replaced with ours; the captured chariots mixed with ours and mounted by our men. The prisoners of war should be kindly treated and kept. This is called "becoming stronger in the course of defeating the enemy."

Know Your Craft

Hence, what is valued in war is a quick victory, not prolonged operations. And, therefore, the general who understands war is the controller of his people's fate and the guarantor of the security of the nation.

Chapter 3

Attack by Stratagem

Win Without Fighting

GENERALLY, IN WAR the best thing of all is to take the enemy's state whole and intact; to ruin it is inferior to this. To capture the enemy's entire army is better than to destroy it; to take intact a battalion, a company, or a five-man squad is better than to destroy them. Hence, to win one hundred victories in one hundred battles is not the acme of skill. To subdue the enemy without fighting is the supreme excellence.

Thus, the best policy in war is to attack the enemy's strategy. The second best way is to disrupt his alliances through diplomatic means. The next best method is to attack his army in the field. The worst policy is to attack walled cities. Attacking cities is the last resort when there is no alternative.

It takes at least three months to make mantlets and shielded vehicles ready and prepare necessary arms and equipments. It takes at least another three months to pile up earthen mounds against the walls. The general unable to control his impatience will order his troops to swarm up the wall like ants, with the result that one-third of them are slain, while the city remains untaken. Such is the calamity of attacking walled cities.

Therefore, those skilled in war subdue the enemy's army without fighting. They capture the enemy's cities without assaulting them and overthrow his state without protracted operations. Their aim must be to take all under heaven intact through strategic superiority. Thus, their troops are not worn out and their triumph will be complete. This is the art of attacking by stratagem.

Attain Strategic Superiority

Consequently, the art of using troops is this:

When ten to the enemy's one, surround him.
When five times his strength, attack him.
If double his strength, engage him.
If equally matched, be capable of dividing him.
If less in number, be capable of defending yourself.
And, if in all respects unfavorable, be capable of eluding him.

Hence, a weak force will eventually fall captive to a strong one if it simply holds ground and conducts a desperate defense.

Beware of "High-Level Dumb"

Now, the general is the bulwark of the state:

If the bulwark is complete at all points, the state will surely be strong.
If the bulwark is defective, the state will certainly be weak.

Now, there are three ways in which a sovereign can bring misfortune upon his army:

1. By ordering an advance while ignorant of the fact that the army cannot go forward, or by ordering a retreat while ignorant of the fact that the army cannot fall back. This is described as "hobbling the army."

2. By interfering with the army's administration without knowledge of the internal affairs of the army. This causes officers and soldiers to be perplexed.
3. By interfering with the direction of fighting while ignorant of the military principle of adaptation to circumstances. This sows doubts and misgivings in the minds of his officers and soldiers.

If the army is confused and suspicious, neighboring rulers will take advantage of this and cause trouble. This is simply bringing anarchy into the army and flinging away victory.

Seek Circumstances That Assure Victory

Thus, there are five points in which victory may be predicted:

1. He who knows when to fight and when not to fight will win.
2. He who understands how to handle both superior and inferior forces will win.
3. He whose ranks are united in purpose will win.
4. He who is well prepared and lies in wait for an enemy who is not well prepared will win.
5. He whose generals are able and not interfered with by the sovereign will win.

It is in these five points that the way to victory is known. Therefore, I say:

Know the enemy and know yourself, and you can fight a hundred battles with no danger of defeat.

When you are ignorant of the enemy but know yourself, your chances of winning and losing are equal.

If ignorant both of your enemy and of yourself, you are sure to be defeated in every battle.

Chapter 4

Disposition of Military Strength

Be Invincible

THE SKILLFUL WARRIORS in ancient times first made themselves invincible and then awaited the enemy's moment of vulnerability. Invincibility depends on oneself, but the enemy's vulnerability on himself. It follows that those skilled in war can make themselves invincible but cannot cause an enemy to be certainly vulnerable. Therefore, it can be said that one may know how to achieve victory but cannot necessarily do so.

Invincibility lies in the defense, the possibility of victory in the attack. Defend yourself when the enemy's strength is abundant, and attack the enemy when it is inadequate. Those who are skilled in defense hide themselves as under the most secret recesses of earth. Those skilled in attack flash forth as from above the topmost heights of heaven. Thus, they are capable both of protecting themselves and of gaining a complete victory.

Win Without Fighting

To foresee a victory no better than ordinary people's foresight is not the acme of excellence. Neither is it the acme of excellence if you win a victory through fierce fighting and the whole empire says, "Well done!" Hence, by analogy, to lift an autumn hare does not signify great strength; to see the sun and moon does not signify good sight; to hear the thunderclap does not signify acute hearing.

In ancient times, those called skilled in war conquered an enemy easily conquered. Consequently, a master of war wins victories without showing his brilliant military success, and without gaining the reputation for wisdom or the merit for valor. He wins his victories without making mistakes. Making no mistakes is what establishes the certainty of victory, for it means that he conquers an enemy already defeated.

Accordingly, a wise commander always ensures that his forces are put in an invincible position, and at the same time will be sure to miss no opportunity to defeat the enemy. It follows that a triumphant army will not fight with the enemy until the victory is assured; while an army destined to defeat will always fight with the opponent first, in the hope that it may win by sheer good luck. The commander adept in war enhances the moral influence and adheres to the laws and regulations. Thus, it is in his power to control success.

Use Information to Focus Resources

Now, the elements of the art of war are first, the measurement of space; second, the estimation of quantities; third, the calculation of figures; fourth, comparisons of strength; and fifth, chances of victory.

Measurements of space are derived from the ground. Quantities derive from measurement, figures from quantities, comparisons from figures, and victory from comparisons.

Therefore, a victorious army is as one *yi* balanced against a grain, and a defeated army is as a grain balanced against one *yi*.

An army superior in strength takes action like the bursting of pent-up waters into a chasm of a thousand fathoms deep. This is what the disposition of military strength means in the actions of war.

Chapter 5

Use of Energy

Build a Sound Organization Structure

GENERALLY, MANAGEMENT OF a large force is the same in principle as the management of a few men: it is a matter of organization. And to direct a large army to fight is the same as to direct a small one: it is a matter of command signs and signals.

Employ Extraordinary Force

That the whole army can sustain the enemy's all-out attack without suffering defeat is due to operations of extraordinary and normal forces. Troops thrown against the enemy as a grindstone against eggs is an example of the strong beating the weak.

Generally, in battle, use the normal force to engage and use the extraordinary to win. Now, to a commander adept at the use of extraordinary forces, his resources are as infinite as the heaven and earth, as inexhaustible as the flow of the running rivers. They end and begin again like the motions of the sun and moon. They die away and then are reborn like the changing of the four seasons.

In battle, there are not more than two kinds of postures—operation of the extraordinary force and operation of the normal force, but their combinations give rise to an endless series of maneuvers. For these two forces are mutually reproductive. It is like moving in a circle, never coming to an end. Who can exhaust the possibilities of their combinations?

Coordinate Momentum and Timing

When torrential water tosses boulders, it is because of its momentum; when the strike of a hawk breaks the body of its prey, it is because of timing. Thus, in battle, a good commander creates a posture releasing an irresistible and overwhelming momentum, and his attack is precisely timed in a quick tempo. The energy is similar to a fully drawn crossbow; the timing, the release of the trigger.

Amid turmoil and tumult of battle, there may be seeming disorder and yet no real disorder in one's own troops. In the midst of confusion and chaos, your troops appear to be milling about in circles, yet it is proof against defeat.

Apparent disorder is born of order; apparent cowardice, of courage; apparent weakness, of strength. Order or disorder depends on organization and direction; courage or cowardice on postures; strength or weakness on dispositions.

Thus, one who is adept at keeping the enemy on the move maintains deceitful appearances, according to which the enemy will act. He lures with something that the enemy is certain to take. By so doing he keeps the enemy on the move and then waits for the right moment to make a sudden ambush with picked troops.

Therefore, a skilled commander sets great store by using the situation to the best advantage and does not make excessive demands on his subordinates. Hence, he is able to select the right men and exploits the situation. He who takes advantage of the situation uses his men in fighting as rolling logs or rocks. It is the nature of logs and rocks to stay stationary on the flat ground and to roll forward on a slope. If four-cornered, they stop; if round-shaped, they roll. Thus, the energy of troops skillfully commanded is just like the momentum of round rocks quickly tumbling down from a mountain thousands of feet in height. This is what "use of energy" means.

Chapter 6

Weakness and Strength

Take the Initiative

GENERALLY, HE WHO occupies the field of battle first and awaits his enemy is at ease; he who arrives later and joins the battle in haste is weary. And, therefore, one skilled in war brings the enemy to the field of battle and is not brought there by him.

One able to make the enemy come of his own accord does so by offering him some advantage. And one able to stop him from coming does so by inflicting damage on him.

Plan Surprise

Thus, when the enemy is at ease, he is able to tire him; when well fed, to starve him; when at rest, to make him move. All these can be done because you appear at points that the enemy must hasten to defend.

That you may march a thousand *li* without tiring yourself is because you travel where there is no enemy.

That you are certain to take what you attack is because you attack a place the enemy does not or cannot protect.

That you are certain of success in holding what you defend is because you defend a place the enemy must hasten to attack.

Therefore, against those skillful in attack, the enemy does not know where to defend, and against the experts in defense, the enemy does not know where to attack.

How subtle and insubstantial that the expert leaves no trace. How divinely mysterious that he is inaudible. Thus, he is master of his enemy's fate.

His offensive will be irresistible if he plunges into the enemy's weak points; he cannot be overtaken when he withdraws if he moves swiftly. Hence, if we wish to fight, the enemy will be compelled to an engagement even though he is safe behind high ramparts and deep ditches. This is because we attack a position he must relieve.

If we do not wish to fight, we can prevent him from engaging us even though the lines of our encampment be merely traced out on the ground. This is because we divert him from going where he wishes.

Gain Relative Superiority

Accordingly, by exposing the enemy's dispositions and remaining invisible ourselves, we can keep our forces concentrated, while the enemy's must be divided. We can form a single united body at one place, while the enemy must scatter his forces at ten places. Thus, it is ten to one when we attack him at one place, which means we are numerically superior. And if we are able to use many to strike few at the selected place, those we deal with will be in dire straits.

The spot where we intend to fight must not be made known. In this way, the enemy must take precautions at many places against the attack. The more places he must guard, the fewer his troops we shall have to face at any given point.

For if he prepares to the front, his rear will be weak; and if to the rear, his front will be fragile. If he strengthens his left, his right will be vulnerable; and if his right gets strengthened, there will be few troops on his left. If he sends reinforcements everywhere, he will be weak everywhere.

Numerical weakness comes from having to prepare against possible attacks; numerical strength from compelling the enemy to make these preparations against us.

Practice Good Intelligence

Therefore, if one knows the place and time of the coming battle, his troops can march a thousand *li* and fight on the field. But if one knows neither the spot nor the time, then one cannot manage to have the left wing help the right wing or the right wing help the left; the forces in the front will be unable to support the rear, and the rear will be unable to reinforce the front. How much more so if the farthest portions of the troop deployments extend tens of *li* in breadth, and even the nearest troops are separated by several *li!*

Although I estimate the troops of Yue as many, of what benefit is this superiority in terms of victory?

Thus, I say that victory can be achieved. For even if the enemy is numerically stronger, we can prevent him from fighting.

Therefore, analyze the enemy's battle plan so as to have a clear understanding of its strong and weak points. Agitate the enemy so as to ascertain his pattern of movement. Lure him into the open so as to find out his vulnerable spots in disposition. Probe him and learn where his strength is abundant and where it is deficient.

Now, the ultimate in disposing one's troops is to conceal them without ascertainable shape. In this way, the most penetrating spies cannot pry nor can the wise lay plans against you.

Be Flexible

Even though we show people the victory gained by using flexible tactics in conformity to the changing situations, they do not comprehend this. People all know the tactics by which we achieved victory, but they do not know how the tactics were applied in the situation to defeat the enemy. Hence, no one victory is gained in the same manner as another. The tactics change in an infinite variety of ways to suit changes in the circumstances.

Now the laws of military operations are like water. The tendency of water is to flow from heights to lowlands. The law of successful operations is to avoid the enemy's strength and strike his weakness. Water changes its course in accordance with the contours of the land. The soldier works out his victory in accordance with the situation of the enemy.

Hence, there are neither fixed postures nor constant tactics in warfare. He who can modify his tactics in accordance with the enemy situation and thereby succeeds in winning may be said to be divine. Of the five elements, none is ever predominant; of the four seasons, none lasts forever; of the days, some are longer and others shorter; and of the moon, it sometimes waxes and sometimes wanes.

孫
子

Chapter 7

Maneuvering

Maneuver to Gain the Advantage

NORMALLY, IN WAR, the general receives his commands from the sovereign. During the process from assembling the troops and mobilizing the people to deploying the army ready for battle, nothing is more difficult than the art of maneuvering for seizing favorable positions beforehand. What is difficult about it is to make the devious route the most direct and to turn disadvantage to advantage. Thus, by forcing the enemy to deviate and slow down his march by luring him with a bait, you may set out after he does and arrive at the battlefield before him. One able to do this shows the knowledge of artifice of deviation.

Thus, both advantage and danger are inherent in maneuvering for an advantageous position. One who sets the entire army in motion with impedimenta to pursue an advantageous position will be too slow to attain it. If he abandons the camp and all the impedimenta to contend for advantage, the baggage and stores will be lost.

It follows that when the army rolls up the armor and sets out speedily, stopping neither day nor night and marching at double speed for a hundred *li* to wrest an advantage, the commander of three divisions will be captured. The vigorous troops will arrive first and the feeble will straggle along behind, so that if this method is used, only one-tenth of the army will arrive. In a forced march of fifty li, the commander of the first and van division will fall, and using this method but half of the army will arrive. In a forced march of thirty li, but two-thirds will arrive. Hence, the army will be lost without baggage train, and it cannot survive without provisions, nor can it last long without sources of supplies.

Deceive Your Opponent

One who is not acquainted with the designs of his neighbors should not enter into alliances with them. Those who do not know the conditions of mountains and forests, hazardous defiles, and marshes and swamps cannot conduct the march of an army. Those who do not use local guides are unable to obtain the advantages of the ground.

Now, war is based on deception. Move when it is advantageous and change tactics by dispersal and concentration of your troops. When campaigning, be swift as the wind; in leisurely march, be majestic as the forest; in raiding and plundering, be fierce as fire; in standing, be firm as the mountains. When hiding, be as unfathomable as things behind the clouds; when moving, fall like a thunderclap. When you plunder the countryside, divide your forces. When you conquer territory, defend strategic points.

Weigh the situation before you move. He who knows the artifice of deviation will be victorious. Such is the art of maneuvering.

Practice the Art of Good Management

The Book of Army Management says, "As the voice cannot be heard in battle, gongs and drums are used. As troops cannot see each other clearly in battle, flags and banners are used." Hence, in night fighting, usually use drums and gongs; in day fighting, banners and flags. Now, these instruments are used to unify the action of the troops. When the troops can be thus united, the brave cannot advance alone, nor can the cowardly retreat. This is the art of directing large masses of troops.

A whole army may be robbed of its spirit, and its commander deprived of his presence of mind. Now, at the beginning of a campaign, the spirit of soldiers is keen; after a certain period of time, it declines; and in the later stage, it may be dwindled to naught. A clever commander, therefore, avoids the enemy when his spirit is keen and attacks him when it is lost. This is the art of attaching importance to moods. In good order, he awaits a disorderly enemy; in serenity, a clamorous one. This is the art of retaining self-possession. Close to the field of battle, he awaits an enemy coming from afar; at rest, he awaits an exhausted enemy; with well-fed troops, he awaits hungry ones. This is the art of husbanding one's strength.

He refrains from intercepting an enemy whose banners are in perfect order, and desists from attacking an army whose formations are in an impressive array. This is the art of assessing circumstances.

Now, the art of employing troops is that when the enemy occupies high ground, do not confront him uphill, and when his back is resting on hills, do not make a frontal attack. When he pretends to flee, do not pursue. Do not attack soldiers whose temper is keen. Do not swallow a bait offered by the enemy. Do not thwart an enemy who is returning homeward. When you surround an army, leave an outlet free. Do not press a desperate enemy too hard. Such is the method of using troops.

Chapter 8

Variation of Tactics

Tactics Vary with the Situation

GENERALLY IN WAR, the general receives his commands from the sovereign, assembles troops, and mobilizes the people. When on grounds hard of access, do not encamp. On grounds intersected with highways, join hands with your allies. Do not linger on critical ground. In encircled ground, resort to stratagem. In desperate ground, fight a last-ditch battle.

There are some roads that must not be followed, some troops that must not be attacked, some cities that must not be assaulted, some ground that must not be contested, and some commands of the sovereign that must not be obeyed. Hence, the general who thoroughly understands the advantages that accompany variation of tactics knows how to employ troops. The general who does not is unable to use the terrain to his advantage even though he is well acquainted with it. In employing the troops for attack, the general who does not understand the variation of tactics will be unable to use them effectively, even if he is familiar with the Five Advantages.

Carefully Consider Advantages and Disadvantages

And for this reason, a wise general in his deliberations must consider both favorable and unfavorable factors. By taking into account the favorable factors, he makes his plan feasible; by taking into account the unfavorable, he may avoid possible disasters.

What can subdue the hostile neighboring rulers is to hit what hurts them most. What can keep them constantly occupied is to make trouble for them, and what can make them rush about is to offer them ostensible allurements.

It is a doctrine of war that we must not rely on the likelihood of the enemy not coming, but on our own readiness to meet him; not on the chance of his not attacking, but on the fact that we have made our position invincible.

Avoid the Faults of Leadership

There are five dangerous faults that may affect a general:

- If reckless, he can be killed.
- If cowardly, he can be captured.
- If quick-tempered, he can be provoked to rage and make a fool of himself.
- If he has too delicate a sense of honor, he is liable to fall into a trap because of an insult.
- If he is of a compassionate nature, he may get bothered and upset.

These are the five serious faults of a general, ruinous to the conduct of war. The ruin of the army and the death of the general are inevitable results of these five dangerous faults. They must be deeply pondered.

Chapter 9

On the March

Occupy Strong Natural Positions

GENERALLY, WHEN AN army takes up a position and sizes up the enemy situation, it should pay attention to the following:

When crossing the mountains, be sure to stay in the neighborhood of valleys; when encamping, select high ground facing the sunny side; when high ground is occupied by the enemy, do not ascend to attack. So much for taking up a position in the mountains.

After crossing a river, you should get far away from it. When an advancing invader crosses a river, do not meet him in midstream. It is advantageous to allow half his force to get across and then strike. If you wish to fight a battle, you should not go to meet the invader near a river that he has to cross. When encamping in the river area, take a position on high ground facing the sun. Do not take a position at the lower reaches of the enemy. This relates to positions near a river.

In crossing salt marshes, your sole concern should be to get over them quickly, without any delay. If you encounter the enemy in a salt marsh, you should take a position close to grass and water with trees to your rear. This has to do with taking up a position in salt marshes.

On level ground, take up an accessible position and deploy your main flanks on high grounds with front lower than the back. This is how to take up a position on level ground. These are principles for encamping in the four situations named. By employing them, the Yellow Emperor conquered his four neighboring sovereigns.

Always Seek the High Ground

Generally, in battle and maneuvering, all armies prefer high ground to low, and sunny places to shady. If an army encamps close to water and grass with adequate supplies, it will be free from countless diseases and this will spell victory. When you come to hills, dikes, or embankments, occupy the sunny side, with your main flank at the back. All these methods are advantageous to the army and can exploit the possibilities the ground offers.

When heavy rain falls in the upper reaches of a river and foaming water descends, do not ford, but wait until it subsides. When encountering "Precipitous Torrents," "Heavenly Wells," "Heavenly Prison," "Heavenly Net," "Heavenly Trap," and "Heavenly Cracks," you must march speedily away from them. Do not approach them. While we keep a distance from them, we should draw the enemy toward them. We face them and cause the enemy to put his back to them.

If in the neighborhood of your camp there are dangerous defiles or ponds and low-lying ground overgrown with aquatic grass and reeds or forested mountains with dense tangled undergrowth, they must be thoroughly searched, for these are possible places where ambushes are laid and spies are hidden.

Make an Estimate of the Situation

When the enemy is close at hand and remains quiet, he is relying on a favorable position. When he challenges battle from afar, he wishes to lure you to advance; when he is on easy ground, he must be in an advantageous position. When the trees are seen to move, it means the enemy is advancing; when many screens have been placed in the undergrowth, it is for the purpose of deception. The rising of birds in their flight is the sign of an ambuscade. Startled beasts indicate that a sudden attack is forthcoming.

Dust spurting upward in high, straight columns indicates the approach of chariots. When it hangs low and is widespread, it betokens that infantry is approaching. When it branches out in different directions, it shows that parties have been sent out to collect firewood. A few clouds of dust moving to and fro signify that the army is camping.

When the enemy's envoys speak in humble terms but the army continues preparations, that means it will advance. When their language is strong and the enemy pretentiously drives forward, these may be signs that he will retreat. When light chariots first go out and take positions on the wings, it is a sign that the enemy is forming for battle. When the enemy is not in dire straits but asks for a truce, he must be plotting. When his troops march speedily and parade in formations, he is expecting to fight a decisive battle on a fixed date. When half his force advances and half retreats, he is attempting to decoy you.

When his troops lean on their weapons, they are famished. When drawers of water drink before carrying it to camp, his troops are suffering from thirst. When the enemy sees an advantage but does not advance to seize it, he is fatigued.

When birds gather above his campsites, they are unoccupied. When at night the enemy's camp is clamorous, it betokens nervousness. If there is disturbance in the camp, the general's authority is weak.

If the banners and flags are shifted about, sedition is afoot. If the officers are angry, it means that men are weary. When the enemy feeds his horses with grain, kills the beasts of burden for food, and packs up the utensils used for drawing water, he shows no intention to return to his tents and is determined to fight to the death.

When the general speaks in a meek and subservient tone to his subordinates, he has lost the support of his men. Too frequent rewards indicate that the general is at the end of his resources; too frequent punishments indicate that he is in dire distress. If the officers at first treat the men violently and later are fearful of them, it shows supreme lack of intelligence.

When envoys are sent with compliments in their mouths, it is a sign that the enemy wishes for a truce.

When the enemy's troops march up angrily and remain facing yours for a long time, neither joining battle nor withdrawing, then the situation demands great vigilance and thorough investigation.

In war, numbers alone confer no advantage. If one does not advance by force recklessly and is able to concentrate his military power through a correct assessment of the enemy situation and enjoys full support of his men, that would suffice. He who lacks foresight and underestimates his enemy will surely be captured by him.

Generate a Fair and Harmonious Relationship

If troops are punished before they have grown attached to you, they will be disobedient. If not obedient, it is difficult to employ them. If troops have become attached to you but discipline is not enforced, you cannot employ them either. Thus, soldiers must be treated in the first instance with humanity but kept under control by iron discipline. In this way, the allegiance of soldiers is assured.

If orders are consistently carried out and the troops are strictly supervised, they will be obedient. If orders are never carried out, they will be disobedient. And the smooth implementation of orders reflects a harmonious relationship between the commander and his troops.

Chapter 10

Terrain

Know the Battlefield

GROUND MAY BE classified according to its nature as accessible, entangling, temporizing, constricted, precipitous, and distant.

Ground that both we and the enemy can traverse with equal ease is called accessible. On such ground, he who first takes high sunny positions and keeps his supply routes unimpeded can fight advantageously.

Ground easy to reach but difficult to exit is called entangling. The nature of this ground is such that if the enemy is unprepared and you sally out, you may defeat him. But if the enemy is prepared for your coming, and you fail to defeat him, then, return being difficult, disadvantages will ensue.

Ground equally disadvantageous for both the enemy and ourselves to enter is called temporizing. The nature of this ground is such that even though the enemy should offer us an attractive bait, it will be advisable not to go forth but march off. When his force is halfway out because of our maneuvering, we can strike him with advantage.

With regard to the constricted ground, if we first occupy it, we must block the narrow passes with strong garrisons and wait for the enemy. Should the enemy first occupy such ground, do not attack him if the pass in his hand is fully garrisoned, but only if it is weakly garrisoned.

With regard to the precipitous ground, if we first occupy it, we must take a position on the sunny heights and await the enemy. If he first occupies such ground, we should march off and not attack him.

When the enemy is situated at a great distance from us, and the terrain where the two armies deploy is similar, it is difficult to provoke battle and unprofitable to engage him.

These are the principles relating to six different types of ground. It is the highest responsibility of the general to inquire into them with the utmost care.

Leaders Must Lead

There are six situations that cause an army to fail. They are: flight, insubordination, fall, collapse, disorganization, and rout. None of these disasters can be attributed to natural and geographical causes, but to the fault of the general.

1. Terrain conditions being equal, if a force attacks one ten times its size, the result is flight.
2. When the soldiers are strong and officers weak, the army is insubordinate.
3. When the officers are valiant and the soldiers ineffective, the army will fall.
4. When the higher officers are angry and insubordinate, and on encountering the enemy, rush to battle on their own account from a feeling of resentment, and the commander-in-chief is ignorant of their abilities, the result is collapse.
5. When the general is incompetent and has little authority, when his troops are mismanaged, when the relationship

between the officers and men is strained, and when the troop formations are slovenly, the result is disorganization.
6. When a general unable to estimate the enemy's strength uses a small force to engage a larger one or weak troops to strike the strong, or fails to select shock troops for the van, the result is rout.

When any of these six situations exists, the army is on the road to defeat. It is the highest responsibility of the general that he examine them carefully.

Know the Situation and Your People

Conformation of the ground is of great assistance in the military operations. It is necessary for a wise general to make correct assessments of the enemy's situation to create conditions leading to victory and to calculate distances and the degree of difficulty of the terrain. He who knows these things and applies them to fighting will definitely win. He who knows them not, and is, therefore, unable to apply them, will definitely lose.

Hence, if, in the light of the prevailing situation, fighting is sure to result in victory, then you may decide to fight even though the sovereign has issued an order not to engage.

If fighting does not stand a good chance of victory, you need not fight even though the sovereign has issued an order to engage.

Hence, the general who advances without coveting fame and retreats without fearing disgrace, whose only purpose is to protect his people and promote the best interests of his sovereign, is the precious jewel of the state.

If a general regards his men as infants, then they will march with him into the deepest valleys. He treats them as his own beloved sons and they will stand by him unto death. If, however, a general is indulgent toward his men but cannot employ them, cherishes them but cannot command them or inflict punishment on them when they violate the regulations, then they may be compared to spoiled children and are useless for any practical purpose.

Know Yourself and Your Opponent

If we know that our troops are capable of striking the enemy but do not know that he is invulnerable to attack, our chance of victory is but half.

If we know that the enemy is vulnerable to attack but do not know that our troops are incapable of striking him, our chance of victory is again but half.

If we know that the enemy can be attacked and that our troops are capable of attacking him but do not realize that the conformation of the ground makes fighting impracticable, our chance of victory is once again but half.

Therefore, those experienced in war moves are never bewildered; when they act, they are never at a loss. Thus, the saying, "Know the enemy and know yourself, and your victory will never be endangered; know the weather and know the ground, and your victory will then be complete."

Chapter 11

The Nine Varieties of Ground

Choose the Battleground

IN RESPECT TO the employment of troops, ground may be classified as dispersive, frontier, key, open, focal, serious, difficult, encircled, and desperate.

When a chieftain is fighting in his own territory, he is in dispersive ground. When he has penetrated into hostile territory, but to no great distance, he is in frontier ground. Ground equally advantageous for us and the enemy to occupy is key ground. Ground equally accessible to both sides is open. Ground contiguous to three other states is focal. He who first gets control of it will gain the support of the majority of neighboring states. When an army has penetrated deep into hostile territory, leaving far behind many enemy cities and towns, it is in serious ground. Mountain forests, rugged steeps, marshes, fens, and all that is hard to traverse fall into the category of difficult ground. Ground to which access is constricted and from which we can retire only by tortuous paths so that a small number of the enemy would suffice to crush a large body of our men is encircled ground. Ground on which the army can avoid annihilation only through a desperate fight without delay is called a desperate one.

And, therefore, do not fight in dispersive ground; do not stop in the frontier borderlands.

Do not attack an enemy who has occupied key ground; in open ground, do not allow your communication to be blocked.

In focal ground, form alliances with neighboring states; in serious ground, gather in plunder.

In difficult ground, press on; in encircled ground, resort to stratagems; and in desperate ground, fight courageously.

Shape Your Opponent's Strategy

In ancient times, those described as skilled in war knew how to make it impossible for the enemy to unite his van and his rear, for his large and small divisions to cooperate, for his officers and men to support each other, and for the higher and lower levels of the enemy to establish contact with each other.

When the enemy's forces were dispersed, they prevented him from assembling them; even when assembled, they managed to throw his forces into disorder. They moved forward when it was advantageous to do so; when not advantageous, they halted.

Should one ask, "How do I cope with a well-ordered enemy host about to attack me?" I reply, "Seize something he cherishes and he will conform to your desires."

Speed is the essence of war. Take advantage of the enemy's not being prepared, make your way by unexpected routes, and attack him where he has taken no precautions.

Victory Is the Only Option

The general principles applicable to an invading force are that the deeper you penetrate into hostile territory, the greater will be the solidarity of your troops, and, thus, the defenders cannot overcome you.

Plunder fertile country to supply your army with plentiful food. Pay attention to the soldiers' well-being and do not fatigue them. Try to keep them in high spirits and conserve their energy. Keep the army moving and devise unfathomable plans.

Throw your soldiers into a position whence there is no escape, and they will choose death over desertion. For if prepared to die, how can the officers and men not exert their uttermost strength to fight? In a desperate situation, they fear nothing; when there is no way out, they stand firm. Deep in a hostile land, they are bound together. If there is no help for it, they will fight hard.

Thus, without waiting to be marshaled, the soldiers will be constantly vigilant; without waiting to be asked, they will do your will; without restrictions, they will be faithful; without giving orders, they can be trusted.

Prohibit superstitious practices and do away with rumors. Then nobody will flee even facing death. Our soldiers have no surplus of wealth, but it is not because they disdain riches; they have no expectation of long life, but it is not because they dislike longevity.

On the day the army is ordered out to battle, your soldiers may weep, those sitting up wetting their garments, and those lying down letting the tears run down their cheeks. But throw them into a situation where there is no escape and they will display the immortal courage of Zhuan Zhu and Cao Kuei.

Troops directed by a skillful general are comparable to the Shuai Ran. The Shuai Ran is a snake found in Mount Heng. Strike at its head, and you will be attacked by its tail; strike at its tail, and you will be attacked by its head; strike at its middle, and you will be attacked by both its head and its tail. Should one ask, "Can troops be made capable of such instantaneous coordination as the Shuai Ran?" I reply, "They can." For the men of Wu and the men of Yue are enemies, yet if they are crossing a river in the same boat and are caught by a storm, they will come to each other's assistance just as the left hand helps the right.

Hence, it is not sufficient to rely upon tethering of the horses and the burying of the chariots. The principle of military administration is to achieve a uniform level of courage. The principle of terrain application is to make the best use of both the high- and the low-lying grounds.

Thus, a skillful general conducts his army just as if he were leading a single man, willy-nilly, by the hand.

It is the business of a general to be quiet and thus ensure depth in deliberation; impartial and upright and, thus, keep a good management.

He should be able to mystify his officers and men by false reports and appearances and, thus, keep them in total ignorance. He changes his arrangements and alters his plans in order to make others unable to see through his strategies. He shifts his campsites and undertakes marches by devious routes so as to make it impossible for others to anticipate his objective.

He orders his troops for a decisive battle on a fixed date and cuts off their return route, as if he kicks away the ladder behind the soldiers when they have climbed up a height. When he leads his army deep into hostile territory, their momentum is trigger-released in battle. He drives his men now in one direction, then in another, like a shepherd driving a flock of sheep, and no one knows where he is going. To assemble the host of his army and bring it into danger—this may be termed the business of the general.

Learn Winning Ways

The different measures appropriate to the nine varieties of ground and the expediency of advance or withdrawal in accordance with circumstances and the fundamental laws of human nature are matters that must be studied carefully by a general.

Generally, when invading a hostile territory, the deeper the troops penetrate, the more cohesive they will be; penetrating only a short way causes dispersion.

When you leave your own country behind and take your army across neighboring territory, you find yourself on critical ground.

When there are means of communication on all four sides, it is focal ground.

When you penetrate deeply into a country, it is serious ground.

When you penetrate but a little way, it is frontier ground.

When you have the enemy's strongholds on your rear, and narrow passes in front, it is encircled ground.

When there is no place of refuge at all, it is desperate ground.

Therefore, in dispersive ground, I would unify the determination of the army. In frontier ground, I would keep my forces closely linked. In key ground, I would hasten up my rear elements. In open ground, I would pay close attention to my defense. In focal ground, I would consolidate my alliances. In serious ground, I would ensure a continuous flow of provisions. In difficult ground, I would press on over the road. In encircled ground, I would block the points of access and egress. In desperate ground, I would make it evident that there is no chance of survival. For it is the nature of soldiers to resist when surrounded, to fight hard when there is no alternative, and to follow commands implicitly when they have fallen into danger.

One ignorant of the designs of neighboring states cannot enter into alliance with them. If ignorant of the conditions of mountains, forests, dangerous defiles, swamps, and marshes, he cannot conduct the march of an army. If he fails to make use of native guides, he cannot gain the advantages of the ground.

An army does not deserve the title of the invincible Army of the Hegemonic King if its commander is ignorant of even one of these nine varieties of ground. Now, when such an invincible army attacks a powerful state, it makes it impossible for the enemy to assemble his forces. It overawes the enemy and prevents his allies from joining him. It follows that one does not need to seek alliances with other neighboring states, nor is there any need to foster the power of other states, but only to pursue one's own strategic designs to overawe his enemy. Then one can take the enemy's cities and overthrow the enemy's state.

Bestow rewards irrespective of customary practice and issue orders irrespective of convention and you can command a whole army as though it were but one man.

Set the troops to their tasks without revealing your designs. When the task is dangerous, do not tell them its advantageous aspect. Throw them into a perilous situation and they will survive; put them in desperate ground and they will live. For when the army is placed in such a situation, it can snatch victory from defeat.

Now, the key to military operations lies in cautiously studying the enemy's designs. Concentrate your forces in the main direction against the enemy and from a distance of a thousand *li* you can kill his general. This is called the ability to achieve one's aim in an artful and ingenious manner.

Therefore, on the day the decision is made to launch war, you should close the passes, destroy the official tallies, and stop the passage of all emissaries. Examine the plan closely in the temple council and make final arrangements.

If the enemy leaves a door open, you must rush in. Seize the place the enemy values without making an appointment for battle with him. Be flexible and decide your line of action according to the situation on the enemy side.

At first, then, exhibit the coyness of a maiden until the enemy gives you an opening; afterward be swift as a running hare, and it will be too late for the enemy to oppose you.

Chapter 12

Attack by Fire

Be Disruptive and Intrusive

THERE ARE FIVE WAYS of attacking with fire. The first is to burn soldiers in their camp; the second, to burn provision and stores; the third, to burn baggage trains; the fourth, to burn arsenals and magazines; and the fifth, to burn the lines of transportation.

To use fire, some medium must be relied upon. Materials for setting fire must always be at hand. There are suitable seasons to attack with fire, and special days for starting a conflagration. The suitable seasons are when the weather is very dry; the special days are those when the moon is in the constellations of the Sieve, the Wall, the Wing, or the Crossbar; for when the moon is in these positions, there are likely to be strong winds all day long.

Now, in attacking with fire, one must respond to the five changing situations: When fire breaks out in the enemy's camp, immediately coordinate your action from without. If there is an outbreak of fire but the enemy's soldiers remain calm, bide your time and do not attack. When the force of the flames has reached its height, follow it up with an attack, if that is practicable; if not, stay where you are. If fires can be raised from outside the enemy's camps, it is not necessary to wait until they are started inside. Attack with fire only when the moment is suitable. If the fire starts from up-wind, do not launch attack from downwind.

When the wind continues blowing during the day, then it is likely to die down at night.

Now, the army must know the five different fire-attack situations and wait for appropriate times.

Those who use fire to assist their attacks can achieve tangible results; those who use inundations can make their attacks more powerful. Water can intercept and isolate an enemy but cannot deprive him of the supplies or equipment.

Consolidate Your Gains

Now, to win battles and capture lands and cities but to fail to consolidate these achievements is ominous and may be described as a waste of resources and time. And, therefore, the enlightened rulers must deliberate upon the plans to go to battle, and good generals carefully execute them.

Exercise Restraint

If not in the interests of the state, do not act. If you are not sure of success, do not use troops. If you are not in danger, do not fight a battle.

A sovereign should not launch a war simply out of anger, nor should a general fight a war simply out of resentment. Take action if it is to your advantage; cancel the action if it is not. An angered man can be happy again, just as a resentful one can feel pleased again, but a state that has perished can never revive, nor can a dead man be brought back to life.

Therefore, with regard to the matter of war, the enlightened ruler is prudent, and the good general is full of caution. Thus, the state is kept secure and the army preserved.

Chapter 13

Employment of Secret Agents

Budget Adequate Funds

GENERALLY, WHEN AN army of 100,000 is raised and dispatched on a distant war, the expenses borne by the people together with the disbursements made by the treasury will amount to 1,000 pieces of gold per day. There will be continuous commotion both at home and abroad; people will be involved with convoys and exhausted from performing transportation services, and 700,000 households will be unable to continue their farmwork.

Hostile armies confront each other for years in order to struggle for victory in a decisive battle; yet if one who begrudges the expenditure of 100 pieces of gold in honors and emoluments remains ignorant of his enemy's situation, he is completely devoid of humanity. Such a man is no leader of the troops; no capable assistant to his sovereign; no master of victory.

Establish an Active Intelligence System

Now, the reason that the enlightened sovereign and the wise general conquer the enemy whenever they move and their achievements surpass those of ordinary men is that they have foreknowledge. This "foreknowledge" cannot be elicited from spirits, nor from gods, nor by analogy with past events, nor by any deductive calculations. It must be obtained from the men who know the enemy situation.

Hence, there are five sorts of spies: native, internal, converted, doomed, and surviving.

When all these five sorts of spies are at work and none knows their method of operation, it would be divinely intricate and constitutes the greatest treasure of a sovereign.

1. Native spies are those we employ from the enemy's country people.
2. Internal spies are enemy officials whom we employ.
3. Converted spies are enemy spies whom we employ.
4. Doomed spies are those of our own spies who are deliberately given false information and told to report it.
5. Surviving spies are those who return from the enemy camp to report information.

Hence, of all those in the army close to the commander, none are more intimate than the spies; of all rewards, none more liberal than those given to spies; of all matters, none are more confidential than those relating to spying operations.

He who is not sage cannot use spies. He who is not humane and generous cannot use spies. And he who is not delicate and subtle cannot get the truth out of them. Truly delicate indeed!

There is no place where espionage is not possible. If plans relating to spying operations are prematurely divulged, the spy and all those to whom he spoke of them should be put to death.

Generally, whether it be armies that you wish to strike, cities that you wish to attack, or individuals whom you wish to assassinate, it is necessary to find out the names of the garrison commander, the aides-de-camp, the ushers, the gatekeepers, and the bodyguards. You must instruct your spies to ascertain these matters in minute detail.

It is essential to seek out enemy spies who have come to conduct espionage against you and bribe them to serve you. Courteously exhort them and give your instructions, then release them back home. Thus, converted spies are recruited and used. It is through the information brought by the converted spies that native and internal spies can be recruited and employed. It is owing to their information, again, that the doomed spies, armed with false information, can be sent to convey it to the enemy. Lastly, it is by their information that the surviving spies can come back and give information as scheduled. The sovereign must have full knowledge of the activities of the five sorts of spies. And to know these depends upon the converted spies. Therefore, it is mandatory that they be treated with the utmost liberality.

In ancient times, the rise of the Shang Dynasty was due to Yi Zhi, who had served under the Xia. Likewise, the rise of the Zhou Dynasty was due to Lu Ya, who had served under the Yin. Therefore, it is only the enlightened sovereign and the wise general who are able to use the most intelligent people as spies and achieve great results. Spying operations are essential in war; upon them the army relies to make its every move.

Translated by Pan Jiabin and Liu Ruixiang
Peoples Republic of China

BOOK TWO
The Art of Execution

Simplicity

Introduction

Simple Ideas, Executed with Gusto, Have a Tendency to Work

GREAT BUSINESS SUCCESSES are most often based on simple ideas:

- Michael Dell revolutionized computers by making them to order.
- Southwest took costs out of the airline business and offered customers lower fares.
- Whole Foods brought a wide selection of organic foods to customers across the country.

Simple ideas, executed with gusto, have a tendency to work. And they work despite their shortcomings.

- Dell computers started out less reliable than store-bought computers.
- Southwest forced customers to travel to less familiar airports.
- Whole Foods' nickname—"whole paycheck"—is almost as familiar as the brand itself.

Executing with simplicity lets you win versus larger competitors. The airline business is being remade by companies with simple business models. Southwest is not alone in offering a focused low-price model. Jet Blue executes that model well, also. Both companies are doing very well against competitors with a more dizzying array of strategies.

United fights back with an expensive (expensive for them, free for flyers) rewards program, an off-price airline named Ted, and their international Star Alliance. They also copy Southwest innovations like boarding by group instead of by seat number. You pick a strategy, they've got it. Complexity built on complexity.

Which airline would you bet on?

Simple ideas win in the marketplace. They win because they:

Lower cost structures. Complexity is expensive. It is expensive to maintain. It is expensive to teach to your employees. It is expensive to explain to customers. More than anything, the intricacies of complexity require costs that are sometimes hidden. Southwest operates one kind of plane—their competitors offer many. Whole Foods is small versus many of their competitors and overall lacks "scale" in its buying—but among organic food producers, it is the big kahuna (and it aggressively uses its relative importance to those particular producers to buy those foods well).

Build mass on a single idea. People can understand simple ideas. It's not that people are simple—it's that people lead complicated and busy lives. In research study after research study, Whole Foods was found to mean "organics." They can be beat on many other measures, but they own that idea in consumers' brains.

Are simple enough that your employees can execute them. Your employees can execute ideas they understand. Your employees' lives are busy too—and their complicated or challenging personal lives will make it difficult for them to focus on their work sometimes. "Work/life balance" will more frequently tilt to an emphasis on their personal life and issues. If you can make it easy for your employees to understand how to succeed at work, and reward them in a way that encourages that behavior, they are more likely to give you what you want. The simpler the idea, the easier it is to get everyone to understand it and make daily decisions that support it.

Think of the changing of the guards of the world's largest companies—from General Motors to Wal-Mart. For years Wal-Mart executed one simple thing—low prices. They built logistics and IT strengths that allowed them to deliver goods to customers at lower prices than anyone else.

Wal-Mart replaced General Motors as the world's largest company. GM operated many divisions, with confusing product lineups. Many of their brands lost their simple distinctiveness (think of the clarity of "Hummer" or "Cadillac" versus "Oldsmobile" or "Buick"). There is little simplicity in the positioning of any GM brand. GM brands that are fuzzy in their meaning—Pontiac and Buick and even Saab—are regularly rumored to be at risk of following Oldsmobile to the brand scrap heap. Today, GM continues to slip in performance.

Advice on simplicity reverberates through history—in the U.S. Army's *Infantry in Battle,* the U.S. Army advises, "In war the simplest way is usually the best way. Direct, simple plans, clear, concise orders, formations that facilitate control, and routes that are unmistakably defined will smooth the way for subordinate elements, minimize the confusion of combat and ordinarily increase the chances of success. In brief, simplicity is the sword with which the capable leader may cut the Gordian knot of many a baffling situation."

IKEA, the chain of mammoth furniture stores, says succinctly, "Simplicity is complexity resolved." IKEA is a mammoth store—but always consistently mammoth. If you have been in one, you know what to expect. You know the layout will be a "racetrack" around the store. You know they sell ready-to-assemble furniture. And you know they sell just furniture and closely related items. Ikea may sell a giant selection of items, but they make that selection very understandable for their shoppers.

IKEA embraces simplicity in their merchandising strategy as well. When IKEA wants you to understand they have chairs for you, they merchandise them—with mass. They put numerous chairs, in a variety of styles and settings, in the front of their stores. Whoever you are, you can't miss that IKEA has a chair for you—and you immediately see how it fits into your lifestyle and your home.

Any merchandising person will tell you simplicity works. In fact, an old merchandising phrase is "Pile it high, and watch them buy." Pile one thing high. A week later, pile up something else and watch that item sell too! Simplicity doesn't limit creativity. It simply channels it.

Committing to a simple idea works.

Chapter 1

Communication

To direct a large army to fight is the same as to direct a small one: it is a matter of command signs and signals.

—Sun Tzu

GREAT COMMUNICATION—winning communication—starts with simplicity. When you deliver a simple message over and over again, you create great clarity with your customers, your employees, and in the marketplace. Thus, a simple, clear message creates an opportunity for your customers or employees to understand, and take action, on the message you are sending.

In advertising, this is called a positioning statement. In a positioning statement, you outline one benefit.

- Volvos are safe.
- Wal-Mart has low prices.
- Target is fashionable.
- FedEx is on time.
- "Nobody doesn't like Sara Lee."

One idea repeated consistently over time has great impact.

Procter & Gamble for years said that Secret antiperspirant was "Strong enough for a man, but made for a woman." Brand manager after brand manager worked with, and executed, that one simple line in a variety of ways. It worked—the Secret brand has done well in the market for many years. After decades of using that same line, how do you evolve it for a new millennium? "Strong enough for a Woman." It's a subtle change and a nice evolution—the simple statement subtly says something about the strength of women. Yet the new line still builds on decades and tens of millions of dollars of investment in the prior line.

It is memorable when an advertising communication works in the marketplace and contains more than a single benefit. The original Miller Lite campaign is one such example. Miller Lite built its business with memorable communications that carried two benefits: "Tastes great, less filling." In the early years, each commercial revolved entirely around those words, with famous athletes arguing over whether the beer was great because it was less filling or it tasted great. The arguments in the commercials told the story of the dual benefits. Repeated over years, with a giant beer industry advertising budget, the campaign worked. And the campaign achieved some fame in the advertising industry precisely because of the multiple benefits of the communication, and the TV ads that effectively delivered that message.

Think of the flip side of simple communications. Companies and brands that change their message annually have no meaning. What's Burger King communicating today? Undoubtedly, something different than they communicated a couple of years ago.

Sometimes, simple communications seem to get lost in corporate gobbledygook. Samsung says, "A better world is our business." NEC is "Empowered by innovation." Fujitsu says, "The possibilities are infinite." Siemens? "Global network of innovation" is their line. Any of these lines could be swapped interchangeably between these companies (and across a variety of additional industries). That's not the simplicity customers look for—good simplicity boils concepts down to their essence. In marketing and branding, taglines that can easily be switched across companies and industries don't reflect that crystallization. Thus, companies need to work even harder and spend more to add relevant meaning to the brand.

Most airlines jump from tagline to tagline. Delta over time has told us "Delta is ready when you are," "You'll love the way we fly," and "We love to fly. And it shows." Each communication may be valid, but failing to execute a *consistent* idea over time deprives your customers of the chance to deeply understand, and identify with, your story.

In politics, the latest term is *framing,* defined by Matt Bai of the *New York Times* in "The Framing Wars" in his July 17, 2005, article as "choosing the language to define a debate and, more importantly, with fitting individual issues into the contexts of broader story lines." In essence, framing is fitting your communication into a relevant context and then telling that story over and over.

Simple Communications to People

When CEO Roger Smith retired from General Motors, he was leaving a company in a difficult condition. His time with the organization included a reorganization of questionable merit, an acquisition of EDS that provided questionable benefits, and a product line of questionable quality. Yet, at the time of his retirement, he felt his greatest failure was in communicating a consistent strategy to his people.

Jack Welch and his disciples do the opposite. They repeatedly sell their strategy to their people. With repetition, they believe, the message gets through to, and eventually gets understood by, the employees of a large organization.

The Sam's Club division of Wal-Mart is a big business whose results have, until recently, lagged its largest rival—Costco. Results improved under Sam's Club president Kevin Turner, who consistently talked to the Sam's team about five priorities. He has said to analysts, "I don't have a different five points for you this afternoon than I did this morning. They are the same five points that we'll have next week and that everyone at every level of our organization is working on." At a recent Wal-Mart annual meeting, CEO Lee Scott said the president of the Sam's Club division had "brought a consistent message to the organization."

Storytelling is an old tradition. Consultant Bob Rosenfeld, of Idea Connections, continually urges his clients to "tell stories." By telling a story, you give people a context in which to understand your communications (a "frame of reference" in advertising terminology). In people terms, storytelling frequently can add a level of sincerity and believability.

Telling a story to people can be a personal experience. It can be a way of connecting a strategy with people in human terms. Norm Rich, the CEO of Weis Markets, says, "You can't sell your ideas until you sell yourself." Use personal stories to sell yourself.

In Summary

Keep it brief.
Say it over and over again.
Make it relevant.

Chapter 2

Timing

When torrential water tosses boulders, it is because of its momentum; when the strike of a hawk breaks the body of its prey, it is because of timing. Thus, in battle, a good commander creates a posture releasing an irresistible and overwhelming momentum, and his attack is precisely timed in quick tempo. The energy is similar to a fully drawn crossbow; the timing, the release of the trigger.

—Sun Tzu

YOU ARE IN CONTROL. That is a simple fact many of us minimize in the marketplace. The projects you choose to work on and the timing you select to launch them are all up to you.

Every company is buffeted by the actions of its competitors. Those competitive actions do affect us—we can't ignore our competitors. However, you can't allow those competitive actions to dominate your every action in the marketplace. If you do, the competitor has already won because it is controlling your agenda. Competitive reactions are just part of the process of business.

Taking control of timing lets you dictate your priorities—and your competitors' priorities. You can't control the competitive marketplace, but you can control the timing of your most important initiatives. Choose to be the competitor in charge of the agenda.

Keep your competitors working on reacting to *you*. Remember the following ideas.

Where your competitor is weak, choose to strike there. A weak store location, a weak line of goods, or a poorly priced set of products all represent opportunities. That weakness may not last forever. When a retailer expands its geography unsuccessfully, look for competitors to move in next to the vulnerable stores. When a company has a narrow line of goods and is successful, look for larger competitors to introduce "flanker" products that compete directly with the smaller competitor. A supplier of mine recently said in a meeting that he was introducing a new line of goods in the coming year specifically to take the sales of a smaller niche player and drive it out of business.

When your competitor stumbles, move in. Look at the airline industry. Southwest's move into the East Coast a few years back created additional troubles for an already struggling US Air. Southwest did not wait for someone else to come in and take US Air's business—it took it itself.

Where you have a temporary advantage, exploit it, and see if you can build long-term advantage from it. Some meat companies own their own livestock (or birds). Some buy on the open market. When the markets are in your favor, your profits are higher than your competitors'. And companies with a cost advantage use some of their additional profits to pick up business at accounts they believe they can keep for the long term.

You can only do a few things well. So execute your most important priorities on the timing that is best for you. Disrupt the marketplace, and your competitors' plans, by being disciplined about bringing your best products and services to market at a timing (and place) of your choice. Bring your best ideas to market just before a key seasonality or with your best customer or with a large customer you would like to pick up.

Good companies make plans and work to execute them. Great companies will also be opportunistic and quickly take advantage of opportunities when they arise—fitting these opportunities into *their* plans as they choose to.

Many well-run retailers alter their growth plans when real estate becomes available. When Kmart sold off significant chunks of real estate, Home Depot moved in and snatched up some new locations. When pressure from Wal-Mart put Northeastern retailers Bradley's and Ames out of business, regional supermarket retailers added many new locations to their portfolios. None of these new locations were in their annual business plans, but plans need to change. And all these real estate acquisitions likely fit a broader framework—the timing simply changed as the opportunities arose.

Move Swiftly

In many Internet businesses, acquiring customers is relatively easy compared to retaining them. A fast-growing, Internet-based deliverer of fresh foods in New York City is FreshDirect. After they acquire a new customer, the hard work begins, and that work needs to be done quickly. Why? Because the speed with which someone makes their first *several* purchases dictates their loyalty. If FreshDirect gets them to make several purchases fast, that customer has changed his or her habit. If the customer makes that switch slowly, he or she may not make that switch at all.

Many businesses find that building loyalty is about moving swiftly. When you acquire new customers, keep them buying from you. "Continuity" programs are designed to do this. Coffee clubs at your local coffee shop are an example of these—buy five cups and your sixth one is free. A strong trial promotion ("first cup of coffee free") combined with a continuity promotion ("buy five cups of coffee, the sixth one is free") is a good way to move customers swiftly into the habit of buying your product.

Keep your programs and timing simple. Complicated plans with complex timing rarely work.

In Summary

Pick the time that works for you.
Be disciplined.
Move swiftly to consolidate gains.

Chapter 3

Clarity

Hence, in night fighting, usually use drums and gongs; in day fighting, banners and flags. Now, these instruments are used to unify action of the troops.

—Sun Tzu

FOR CUSTOMERS, a strong brand has great clarity in its meaning. Consistent execution, repeated over years, brings clarity to a brand. Some examples:

- Hertz has executed a high-service strategy for years. Hertz Gold service levels and ad campaigns like "Not Exactly" (if it's another rental company, it's not exactly Hertz, not quite as good as Hertz) have given the brand clear service leadership. Maybe not whom you call first for the best price, but the place you know will give you the best service.
- Apple has consistently innovated in computers and electronics. Their brand isn't always the sales leader, but it is the perceived innovator. This strength allowed the brand to jump into digital music, and the company to shift its sales mix significantly into this new business.
- Godiva sells upscale chocolates. With premium packaging and premium prices, it has resisted cheaper offerings. Godiva remains a vogue choice for gift giving. Have you ever seen Godiva displayed in a low-end retailer? In general,

> you will find Godiva in higher-rent districts, in settings that reinforce the fashion mystique of the brand. New York's new Time-Warner Center has very pricey real estate—and you will find a Godiva chocolate shop there. When Godiva seeks distribution outside of its own stores, it is very selective. Godiva is found in department stores, a few specialty shops, and a few upscale food retailers.

Clarity of distribution reinforces clarity of a brand. Manufacturers who proactively manage their distribution generally end up with retailers who add value to their brand. Godiva is found in settings that reinforce the brand's fashion imagery. The premium settings for Godiva strengthen the meaning of the brand.

Years ago, Coors had a mystique fueled by its distribution. For most of its history, Coors's distribution was restricted to eleven states near its Colorado home. Travelers coming home from business trips to that geography would bring home a six-pack of Coors. Stories grew up (some false, but still complimentary to the brand) about why Coors was available only in a limited geography. Sometimes the story was that Coors could travel only so far because of the way it was made or because of the refrigerated trucks Coors had to be transported on. Or Coors was just so popular in its home geography that there wasn't enough for the whole country. The 1977 movie *Smokey and the Bandit* centered on a shipment of Coors from Texas to Georgia. In fact, the product was not pasteurized and so needed to be kept cold, increasing the complexity and cost of distribution.

When they expanded nationally, that mystique was lost. Two breweries were opened outside of the West. And a tagline of fifty years, "Brewed with pure Rocky Mountain spring water," had to change. The product itself was forced to stand on its own merits—taste. While Coors is good, it is not necessarily better than any other beer. Today, the Coors mystique has been so diminished that it now lags its Coors Light "flanker" in sales. Regional "craft" beers (Kona Brewing in Hawaii, Brooklyn Beer in New York, and many others) now own the mystique of the regionally available beer.

Clarity Through People

Plans need to be executed by an organization's people. Guiding your people to good execution requires clarity in how you communicate to them. Bring clarity of your company's vision, or your company's desired meaning, to customers by:

Building Core Values. Building a consistent set of expectations can drive desired performance across levels of management. At Costco, the culture encourages getting ever more value for its "members." Stories are told about products, like salmon, where the company was able to get even greater value for its members—lowering prices and increasing quality over many years as Costco grew. Building value for members drives actions of employees across the organization. Anyone at Costco knows that adding value to members is what that company is about.

Making "heroes" of employees. Make heroes of employees who bring your values to life. Skillful communicators are constantly involving their audience. Skillful leaders constantly involve their own people as examples of behaviors they want repeated. Employee-of-the-month-style programs can build this—particularly when simple criteria for selection are communicated. These programs celebrate behaviors that you want replicated throughout the organization.

Celebrating successes. The successes you choose to celebrate will build an expectation of the behavior that gets rewarded. If you want to build a culture of customer service, build some customer service measures—and publicly celebrate when those measures are reached. Then go set some more. Try to set measures that can be accomplished but still reflect sincere progress.

Trader Joe's president Doug Rauch said of managing his workforce (a relatively high-turnover retail workforce) at a January 2006 food conference, "There's no magic to it other than keeping it simple and continuing to execute."

In Summary

Synchronize your actions to get clarity.
Set expectations.
Reinforce behavior you want to see again.

Chapter 4

Inspiration

In war, numbers alone confer no advantage.

—Sun Tzu

THROUGHOUT HISTORY, smashing victories have been won by people with fewer resources. Look at examples from military history such as Horatio Nelson's victory at Trafalgar against a more numerous French and Spanish fleet, by an American Revolutionary army that was poorly trained by the standards of the day, and by successful guerilla movements across the globe that, when started, were significantly undermanned in comparison to their more organized opponents—whom they eventually beat.

The winners in these examples all had some intangible advantages on their side:

Trust in a leader. Nelson was clearly the most brilliant naval mind of his day. With a track record of significant victories, his unconventional plan at Trafalgar was faithfully executed by ships full of people who believed in his leadership.

New tactics or technology. Faster, better adoption of new tactics or technology can bring significant advantage. The longbow allowed a smaller British force to defeat a much larger French force in the 1300s. Outnumbered four to one at the battle of

Crecy, the British, armed with the longbow, won handily. Today, adoption of Internet solutions is allowing many smaller companies to achieve market success against much larger and better-financed rivals.

Belief in a cause. Successful revolutionary movements generally articulate a belief in an alternative vision, a cause to rally around—a set of words or beliefs that come to have meaning.

Help your employees find the inherent value of their work. Most work serves people—it serves your customers. Products and services get used by people who find value in them. Profitable enterprises provide good livings for the families these companies support.

Helping your employees feel the inherent value, or the "higher calling," in their work builds loyalty to your company and a willingness to sacrifice a little to make your company successful. There are many organizations that people do not join for pay or status or advancement—religious, teaching, and charitable professions are examples. These companies are filled with highly loyal people whose motivation and allegiance to the organization, or the principles of the organization, have greater depth. Why can't some portion of that motivation be harnessed to your company?

Helping your employees find the greater good in their work is also inherently motivating. Consultant Ian Lazarus writes about a job he had with a health-care call center: "I arrived at work one morning to learn that one of our nurses, upon taking a call from a diabetic patient, recognized the risk of shock and persuaded family members to take him to the hospital. The ER doctor confirmed the patient was approaching diabetic shock and would have died without timely treatment."

Teddy Roosevelt said, "Far and away the best prize that life has to offer is the chance to work hard at worthwhile work." Help your employees feel their work is worthwhile. Find inspiration in the following.

Customer satisfaction with the work. Remind your employees of how their work is valued by their customers. Use customer testimonials to connect people's work with the value your customers find in that work.

Being of service to others. When I got into retailing, a mentor of mine advised, "Retailing isn't a job, it's a lifestyle" (because retail stores are open seven days a week, and in some cases twenty-four hours a day). Over the years, my kids have asked me, "Why do you have to go to work tomorrow?" (or "tonight" or "this weekend"). I have become comfortable answering, "Because people are counting on me." It is true, and it reflects the meaning of work I want my kids, and employees, to embrace.

What it means to your family. Some people I have worked with have expressed their loyalty to a company by talking about how a company, and the paycheck that comes from that company, "puts bread on our table." With the right sincerity, it's an expression that puts a positive, loyalty-oriented meaning on the need to work for a living.

Communicate a Common Inspiration

Building consistent execution across a company can be tough. We all want every individual to be using his or her knowledge to build a company in a common direction. We want people's commitment, so we want some amount of their personal creativity in the company. That personal creativity, that ability to do their job their way, is what makes them psychologically invested, or "bought in," to your company and its goals. But you need that creativity to be channeled in a specific direction. It must be focused on a consistent vision for the company. Otherwise, costs can get out of control and the actions of your employees will be random and uncoordinated with larger strategic objectives. A little common inspiration can drive common focus to people's execution.

Managing is an inexact science. Much of the communication that goes out to an organization's employees is never even understood. Many companies have a mission, a vision, values, *and* objectives—whew! The CEO and executive managers read these—but few, if any, companies can get that volume of corporate word-ology understood by their employees. It just becomes noise that distracts them from their day-to-day work. Corporate strategies that can be understood at an emotional level are much more likely to be understood and followed.

Stylish Wegmans food markets take their inspiration from Europe. Their food is European-inspired—with French patisserie shops and European brick ovens in-store. Their newer stores have a façade like a European village; some even have a clock tower. European inspiration drives a consistency of execution and supports an internal language about food and the shopping experience.

The president of Haggen's Supermarkets in the Northwest states simply, "Food is the fashion business." This simple statement gives their merchandising and marketing departments permission to change and innovate. Would an employee of this company expect to be on the leading edge of food trends—of course!

IKEA defines its inspiration as "people's lives at home." Simple and clear—it tells everyone in the IKEA organization where the company's forces will be concentrated.

Inspiration like this is shareable. It doesn't tell you exactly what to do, but it does give a group of people common notions about what to build toward.

In Summary

Choose where to be strong.
Utilize unconventional tactics.
Build strong beliefs.

Chapter 5

Leadership

When the general is incompetent and has little authority, when his troops are mismanaged, when the relationship between the officers and men is strained, and when the troop formations are slovenly, the result is disorganization.

—Sun Tzu

WHEN WAL-MART, the world's biggest (and perhaps most feared—by its competitors) retailer, entered the UK retailing market with the purchase of Asda, many thought they knew what the outcome would be: Wal-Mart would dominate the market. After all, Wal-Mart has great buying clout, well-developed systems, and a very good track record. Tesco, one of the United Kingdom's largest retailers, had other ideas. "We decided that instead of waiting to see what they would do and respond, we at Tesco would set the agenda for Wal-Mart," said Sir Terry Leahy.

Tesco set the agenda for Wal-Mart based on price. In its 1992 Annual Report, it stated simply, "We sell for less." An unlikely strategy against the world's largest company, but one that Tesco thought it had the cost structure to support. It also had service, as Tesco is the world's largest Internet-ordering service for grocery delivery. They also place a high value on in-store service. Tesco thought it had a winning hand.

And Tesco achieved strong success—the number one market share in the UK food business. In 2005 Wal-Mart's international president said, "Our biggest challenge is Asda in the UK. Total market growth in the UK has been declining this year. If you follow the UK market, you know that profit warnings are everywhere. There's been one winner so far and that's Tesco."

David always has a chance against Goliath, if he can execute his plan.

GM used to be the world's largest company. But GM continues to slide and lose market share and cut jobs. A culture that lacked leadership was partly to blame during GM's slide in the 1970s, '80s, and '90s.

John DeLorean, the GM executive who left to start his own car company, told of a meeting where a minor point of compensation was being discussed. Suddenly, then-chairman Richard C. Gersternberg says, "'We can't make a decision on this now . . . I think we ought to form a task force to look into this and come back with a report in 90 to 120 days. Then we can make a decision.' He then rattled off members of the task force he was appointing. The whole room was bewildered but no one had the courage to say why. Finally Harold G. Warner, the snow-white-haired, kindly executive vice president, who was soon to retire, broke the silence. 'Dick, this presentation is the result of the task force you appointed some time ago. Most of the people you just appointed to the new task force are on the old one.'"

Leaders accept risk. They work to understand the risks they are taking, to manage those risks, and to reduce them.

Lead by Listening

Procter & Gamble's CEO A. G. Laffley came up through the P&G ranks. Like many successful leaders, listening takes an important role in his leadership. Laffley's listening is often targeted toward customers. Focus groups, one-on-ones, and various forms of research are one of P&G's most basic stocks in trade.

A. G. Laffley, like many at (or trained by) P&G, makes listening to customers a foundation of its management. In a recent article, he said, "The simple principle is find out what she [the customer] wants and give it to her."

Leading companies do this repeatedly—they find out what their customers want, and they give it to them. Leaders listen:

Formally. They have structured feedback loops that regularly give customer, employee, or market feedback. These provide highly organized feedback like monthly tracking surveys that ask the same set of questions of customers month after month. Changes in response to these surveys can frequently be tied back to organizational, or competitive, changes. This kind of feedback is very easy to share with peers and subordinates. Accordingly, it can frequently be used to drive actions and improvements.

To knowledgeable industry experts. It helps to have a network of people you can rely on for advice. The best resources are people who know some aspect of your business deeply that is different from your personal experience or knowledge. If your specialization is classic brand marketing—seek out and learn from people who know direct marketing.

To the man on the street. Talking with people who use your business, or a competitor's, also provides informal feedback. This feedback may be anecdotal but it adds to a leader's breadth of knowledge about his or her business. You shouldn't take anecdotal information at face value, but it does provide helpful clues for further investigation and formalized research.

To employees. A customer's products or services need to be executed through employees. If the employees feel they are being listened to, they will have a greater buy-in and a greater mental/

emotional stake in the business. They frequently are among your company's strongest backers and are out talking to their friends and neighbors about your company. Use their ideas.

In Summary

Set your own agenda.
Be focused.
Stay in touch through constant listening.

Create Alignment

Introduction

Good People, Working Toward a Common Goal

"When aligned around shared values and united in a common vision . . . ordinary people accomplish extraordinary things."

—Ken Blanchard

GOOD PEOPLE, working toward a common goal, can accomplish virtually anything they set their mind to. The ancient Greeks were a fearsome fighting force because they arranged themselves in a formation called a phalanx. This fighting structure allowed each Greek to support his neighbor in battle. Centuries later, the Scots won many battles against the English with their tight formations and coordinated tactics. In business, too, alignment builds strength.

Much alignment can be built through the basics of:

- Setting objectives
- Measuring performance against those objectives

Setting objectives communicates to your employees what is important and what should be prioritized in your organization. When this is done well, your people will unite their individual objectives for personal performance and recognition to larger company objectives.

Build Alignment, and Empower Individuals

Few employees get up each morning intending to do a bad job at work. Most every employee gets up wanting to do what is right for the organization. They come to work looking to do their job—and looking to do it well.

As a former senior leader of a company once rated number one on *Fortune*'s 100 Best Companies to Work For list, I know that hourly employees are much like those in management roles. No manager wants to be told exactly what to do and how to do it. We all want to be able to bring a little bit of ourselves to the job. We want to be able to figure out how to solve a problem, and to utilize our own creativity and judgment in doing so.

Everybody wants to feel good about the work he or she does. But few organizations empower hourly workers or lower-level employees to make decisions. To a large extent, most companies don't know how to give workers the right amount of latitude. Empowering employees means that management must take a different role, and that teams must change the way they work together.

Evaluate Quality of Thought, Not Quality of the Result

If people are going to have the freedom to bring a little bit of themselves to a job, then they are going to execute the job differently than you might. How do you evaluate that work when it is done differently than you would do it? You have to look at what people were thinking when they made their decisions.

When I worked in brick-and-mortar retailing, I would travel to different stores and evaluate the quality of their in-store displays. I could visit each store only infrequently, and I wanted to ensure that good decisions would be made by the in-store staff on the many days I was not able to be there.

When I saw something that was wrong, I focused not on changing it, but instead on changing the thinking that had created it. By asking questions, rather than making statements, I gained information. I asked questions such as, "What are you thinking here?" "Why did you decide to do it this way?" Sometimes the employee's thinking was exactly right (though the result was off), and I would understand how he or she got to where he or she was—and learn something new in the process. At other times I would find out where his or her thinking or assumptions had gone astray and correct it for the future. If the thinking behind the decisions is right, then the right things will have a tendency to happen even when you are not around to manage them directly.

Guidelines Are Not Rules

Guidelines provide principles for decision making. They provide the framework for making the right decisions in changing circumstances. Building guidelines into your businesses lets your people make the right daily or hourly decisions in alignment with broader goals.

Wegmans Food Markets published key guidelines for operating their in-store sub shops. The guidelines gave employees at the store level the latitude to make good decisions. They also set limits on where employees shouldn't have decision-making authority (like which breads to use) and inherently gave permission to make decisions in other areas (like the hours of each sub shop). These guidelines maintained the consistency of the service experience and of the brand. The guidelines enabled the corporate-level managers to maintain control over key aspects of the product and service experience, ensuring a consistent product and experience. Those guidelines also gave employees at the store level the opportunity to make decisions that best serve the day-to-day running of the store.

Gain Buy-In for Objectives

Setting objectives then measuring against them is more effective if the people accountable for the objectives truly buy into or support them. Discern where your goals and objectives match those of people around you. This is known as "safe ground." From these common objectives, you can start negotiating or discussing the issues at hand. Create wins here, at the intersection of everyone's objectives. You will find that creating wins on safe ground creates new safe ground. The next set of objectives you set will also be readily attainable—and then the next, and then the next, as you continue to build buy-in. As you progress, you will find ideas that previously might not have been supported are now safe ground.

A friend of mine describes this as "supporting someone else's objective." Find someone whose objectives mirror your own and lend them your support. You have discretion over what you decide to support and how much of your time you choose to give. However, since you are supporting an objective that someone else is already invested in, the buy-in will already be in place. Few important accomplishments can be done by ourselves. Aligning people to our goals allows each of us to achieve larger accomplishments.

Chapter 6

Mass

We can form a single united body at one place, while the enemy must scatter his forces at ten places.

—Sun Tzu

BEFORE WORLD WAR II, the U.S. Army gave infantrymen the following instruction: "Generalship consists of being stronger at the decisive point—of having three men to attack one. If we attempt to spread out so as to be uniformly strong everywhere, we shall end by being weak everywhere. To have a *real* main effort—and every attack and every attacking unit should have one—we must be prepared to risk extreme weakness elsewhere."

The end result of successfully building alignment is to get as much mass at the critical point as possible. You do not want to place mass (be it in troops, supplies) randomly. You want mass where it will do you, and your organization, the most good. History has taught us that the battle doesn't always go to the side with the biggest army. You want to be "bigger" at the decisive point—not at points that don't matter very much.

For example, Wal-Mart wins every time on price—they have made that their decisive point. They have built an organization that wins on low-cost delivery of goods to a store. That is a *consumer-relevant* point—customers care about price. Wal-Mart has utilized their mass where it counts for the customer.

Disney's theme parks beat their competitors every time on service. Spend half a day at Universal Studios in Orlando, Florida, and half a day at Disney World, also in Orlando. The service difference is clear. Disney runs on time and its employees are better trained. You may have to wait in lines, of course, but Disney executes the people processes around those lines well. Disney's people know how to answer your questions. Disney's staff seems to be well organized. Even Disney's internal bus system has a high on-time rate. Universal, by comparison, just doesn't execute with the same dependability. Waiting times at Universal's lines are unpredictable. Hourly service personnel lack the attention to customer service you feel at Disney. Universal has some faster rides and bigger thrills (for Universal, this may be a decisive point *they* win on for teens)—but the dependability of the experience at Disney is demonstrably more consistent and better. For many families, that's a decisive point. Particularly when backed up by the obvious creativity and storytelling Disney brings to the experience.

Having mass at the decisive point doesn't excuse extreme weakness elsewhere. The Yugo automobile won on price, but that low price was delivered at the expense of reasonable quality. Webvan won on service with many customers during its brief existence, but it never had a solid economic model.

In a competent organization, with a reasonable business plan, creating mass at a decisive point:

- Wins over customers—because they know *something* to love about you.
- Builds good profitability—because customers who choose to shop with you will be willing to pay fairly for the advantages you deliver.

Inherently, where you build mass must matter to enough customers that they are motivated to buy more of your products and services or to recommend your products to other potential customers.

A few years ago, a food product that featured low carbs mattered. That was, briefly, a very decisive point to win on. But that has faded significantly. Laptop computers used to compete aggressively on size and weight. But as more and more reasonably priced lightweight laptops have become available, manufacturers shifted to compete on other features (like quality of the flat-screen picture). Over time, make sure the decisive point you compete on is decisive with your customers.

Coordinate and Communicate

Consumer marketers use repetition in their communications. One metric they look for is frequency. They want each target customer to see the same message several times. They know repetition motivates action. It takes a while for any message to make enough impact to motivate customers to action, that is, buying a product or service.

When introducing a new product, consumer packaged goods companies spend extensively on marketing and promotions in the first few months of the life of that product. This is the critical point. Lots of money is spent getting a new product into distribution. The marketing and promotion creates product movement that drives sales and maintains distribution. Expenditures against other products/brands from that company may suffer in that year to ensure there is sufficient spending to make the new product a success.

When Gillette introduced its six-blade razor, it even bought a Super Bowl advertisement *before* the new product was fully available. This is similar to how movies are advertised in advance of their distribution, to ensure people show up at the movie when it premieres, because there are always plenty of movies that can fill up a theater and knock out an underperformer.

Marketers build mass across all of their communications by executing a single, tactical idea well. UPS is "brown"—in its planes, trucks, and the dress of its people. Target's bull's-eye logo is used in all forms of its advertising and in its store designs. For many years, gifts from Tiffany's have come in a soft blue box. It's now become a distinctive part of Tiffany's branding, and the receiver knows what retailer the gift is from before they even open the gift.

As you go to market, overwhelm your customers. Pick the points where you will have greater mass, greater focus, than your competitors. That could be a point in time—like when you introduce a new product. It could be a particular feature of your product or service—like price. After you pick where you will create advantage, work hard to win at that decisive point.

In Summary

It's not how much you spend—it is when you spend it.
Identify critical points.
Don't just win at the critical point—win decisively.

Chapter 7

Fight Your Most Important Fight

In encircled ground, I would block the points of access and egress.
—Sun Tzu

LIMITING A COMPETITOR'S ability to compete with you is a common practice in business. It forces your competitors to compete with businesses other than your own. Limiting a competitor's access is commonly practiced in the real estate industry by using a couple of different methods:

Restricted-use agreements. When a supermarket exits a popular strip mall, it commonly signs an agreement that prohibits the releasing of that spot to another supermarket. Those agreements are not free. However, when you consider the fact that most supermarkets vacate their current location because they are relocating nearby, preventing the opening of a new competitor in their own backyard is worth the investment.

Keeping open a poorly performing location. Retailers frequently allow a poorly performing location to remain open in order to keep a competitor from coming in. The company feels that the risk of losing customers to a competitor that might take over that location outweighs the cost of keeping open the underperforming location.

Secondary branding. Some retail chains operate secondary brands that keep competitors out of strip mall locations. These secondary brands don't carry the company's flagship brand but instead compete under another, less well known and less advertised brand name in the same category. These secondary brands are not as strong as the flagship brands—sales per location are lower. But the secondary brands lock up a certain amount of locations away from their competitors. Better to run a secondary brand, which doesn't cannibalize your primary brand, at breakeven than to allow the growth of a strong competitor in those locations and disrupt your flagship brand.

Buying up land. In some developed, populous real estate markets, there is only so much land available that is zoned for particular uses. Buying up land that is available can effectively lock a competitor out of a geographic region. This is a strategy a variety of retailers have used against retailing giant Wal-Mart. In many communities, local supermarkets have bought land near their locations. They may or may not develop that land someday—but in the meantime, it ensures that Wal-Mart, or other competing retailers, can't open a new superstore next door to them. This tactic keeps their competition farther away from their own profitable operations.

All of these techniques involve spending money to avoid a fight with a competitor. Obviously, these companies believe that the money spent to avoid a fight is less money than the cost of the fight itself. Or as Sun Tzu says, *Neither is it the acme of excellence if you win a victory through fierce fighting.*

The Pareto Mentality

A Pareto chart ranks a problem from its most frequent instance to its least frequent instance. A common use for a Pareto chart is to rank defects in a manufacturing operation. Generally constructed as a bar chart, it gives a visual picture of a problem. The most frequent defects stand out—they will be the highest bars on the bar chart. Conversations, and work, then go toward fixing the problems that occur most frequently, now that they have been identified. Pareto charts help focus an organization on its most important work.

As an example, say you want to look at reducing the amount of defects in manufacturing operations. A series of Pareto charts can define the problem—and make sure your work is organized to address it. What are the most frequent defects? What are the defects customers complain about the most? What are the problems that cost the most to fix? A couple of Pareto charts that address those questions can help define where problems reside in your organization and its processes and where to focus first.

Pareto charts help create alignment because they bring facts to a situation in both an impartial and an actionable way. Complaints are one example of this. Frequently complaints have an emotional aspect to them. Perhaps a particular complaint goes directly to the president, or a particularly angry customer's complaint is circulated. Those may not be the most important problems to focus on fixing. They may instead be issues that have gotten unusual visibility. Pareto charts take the emotion out of addressing a problem because the issues are arranged in a factual and unbiased way. Using Pareto charts helps you get past emotions and tread-worn conversations and can enable you to achieve a clearer focus on important, granular pieces of a particular issue. Organizations that use this tool frequently are said to have a "Pareto mentality"—they are organizations focused on improving their most important problems. They know they are working on the most important problems because they are continually prioritizing their issues—continually doing Pareto charts.

Be Finicky

For most companies and brands, what you *don't* do best defines your efforts. Coach doesn't sell cheap handbags. Celestial Seasonings tea doesn't contain artificial ingredients. The products and customers they *don't* pursue say as much about who they are as the products and markets they *do* cater to.

In essence, companies that are focused are finicky. They are finicky about what they do and how it is executed.

Be finicky about costs. No company can afford to do everything. Really good companies are particular about the costs they add. Any airline is eager to get new customers into its frequent-flyer program—it builds loyalty and frequency of higher-margin business travelers. Thus, airlines will work hard to get you signed up.

Not JetBlue. Want to register for the JetBlue frequent-flyer program? Register online. At the airport, waiting to get on a flight? Register online. Why would JetBlue take a chance on losing the enrollment, and future revenue, of potential high-volume frequent fliers by turning them away at a point of contact to do business online? Because they know that a discount carrier needs a disciplined approach to costs—and the Internet is a frugal way to manage such a program because there is little employee training involved, and your customers enter and manage their own information.

Be finicky about your brand. Whole Foods is known for organic and natural foods. The company's CEO, John Mackey, talks about being a socially responsible company. In 2005, Whole Foods announced it was not going to sell live lobsters unless a more humane way to hold and transport live lobsters was developed (it would, however, continue to sell frozen dead ones). This move seemed a bit quirky to industry insiders. But it also underscored the core beliefs of the company and the meaning of their brand.

Mid-quality retailer Trader Joe's exhibits a similar finickiness. Trader Joe's is known for its interesting assortment of packaged goods—almost all sold under its own brand name at very good prices. In fact, if the product can't be packaged under its own brand name, you might not find it there at all. Want Coke or Pepsi? Go somewhere else. Both Whole Foods and Trader Joe's do an outstanding job with their particular niche—and they are demonstrating that niches can be big businesses.

Be finicky about your business model. Increasingly, companies are just focusing on what they do best. Remember the conglomerates of the 1980s, with disparate companies linked only by a common holding company? Some examples: AMF made both bowling equipment and Harley-Davidson motorcycles; Sara Lee Corporation made pound cake, Hanes underwear, and Coach leather bags. These holding companies are generally gone or broken up. Harley-Davidson and Coach leather are now their own independent companies. Often, the dismembered companies do much better on their own. What you refuse to do allows you to focus your resources on your most important fights. And win.

In Summary

Put your resources where they will have greatest leverage.
Use Pareto thinking to build focus.
Just say no. Frequently.

Chapter 8

Coordinate Resources

Generally, management of a large force is the same in principle as the management of a few men: it is a matter of organization. And to direct a large army to fight is the same as to direct a small one: it is a matter of command signs and signals.

—Sun Tzu

TOYOTA PRESIDENT Fujio Cho once remarked that, "Detroit people are far more talented than people at Toyota. But we take averagely talented people and make them work as spectacular teams." Is there a science to teamwork? How do you create the alignment that leads to spectacular teamwork?

Bottom Up versus Top Down

Teamwork comes, in part, from team members who know their opinion counts. Organizations that have a strong team culture often give decision-making power to those people who are closest to the decision. This is often referred to as "delegating to the lowest possible level in an organization." If someone underneath you can make a capable decision, then that is who should make it.

Be proactive. If you, instead of your boss, can make a decision in your organization, make it. If you are a manager, empower your people to do so. Maurice of Saxony, the sixteenth-century German military and political leader, writes, "Few orders are best, but they should be followed up with care." More orders don't bring clarity, they bring confusion. Or, as Maurice also writes, "Many generals believe that they have done everything as soon as they have issued orders, and they issue a great deal because they find many abuses." Identify problems when you find them, of course. But keep focused—issuing a few important instructions is much preferable to many orders. People who work with you will better understand their priorities—and your instructions will be treated with a greater importance—if you issue few of them.

Additionally, work to keep decision making at the lowest possible level as you issue instructions. If subordinates will identify the areas to be worked on, and issue orders accordingly, that is preferable to the next level-higher supervisor's issuing the exact same instructions. The buy-in to performing the work will be greater. The same is true for your peers—the more you can help them identify and prioritize the issues in their organizations on their own initiative, and issue their own instructions to fix those issues, the better, and faster, those issues will be fixed.

Customer Centricity

Let the customer be your lightning rod. Decisions made without the customer in mind have a tendency to not stand the test of time. What costs do you add to your organization? What innovations do you bring to market? Are these innovations and the costs associated with them adding value for the customer? Are you spending money where your *customers* would choose for you to spend that money? For example, your organization may think that new fancy packaging is appealing, but does it serve customers who have gotten familiar with your brand and don't want to hunt for it on the store shelf?

Customer-driven organizations ask customers what they are looking for. And make decisions accordingly. We all have ideas on what our customers want us to do. But organizations that add costs and features that customers don't want, or don't understand, aren't investing for the long-term health of their business.

There are two factors that have to come together for true customer centricity:

- Customers have to want your offering. Your product or service has to fill a need.
- Customers have to understand your offering. They won't connect with something they don't understand.

I had a presentation from a food company that wanted me to carry its premium-priced products. It told me its products were:

- All natural
- Made with a proprietary process
- Made from raw materials whose genetics it controlled
- Personally endorsed by the leading expert in that field
- Produced in small quantities so it could control the quality

Whew! No wonder its product was premium priced—and it was struggling to get the premium for it in the marketplace. You won't be surprised to know that the company goes from crisis to crisis because of high prices and low volume. It had more stories to tell about its product than any one customer could understand.

Know Your Position

A great basketball team doesn't send out three centers and two guards to play a game. No football team plays three quarterbacks at the same time on offense. The roles are defined, and people know what role they are supposed to play in the game. That works for the team. In business, make sure each member of your team knows what his or her role is. What does success in each person's position look like? The better each is able to describe it, the more each is able to achieve it.

Steve Jobs and Steve Wozniak made a natural team inventing the first ready-made personal computer. They had different skills. Wozniak's talents were in engineering; Jobs had great creative and marketing instincts. Together their combined skills turned Apple into a billion-dollar company.

Share Common Sacrifices

Most every company or industry has some work the business requires that is out of the norm. If you have a business in a summer resort area, extra hours will be required in the summer. If you are a seller of garments, then you may work 100 hours during "fashion week" so you can spend as much time as you can with your buyers. Sharing these sacrifices around the organization—and getting the entire organization in tune with these important business needs—builds a culture of teamwork.

One HR executive I knew joined a fashion retailer and created alignment in the organization around the seasonal needs of the business. It is common among fashion retailers to "go black" from Thanksgiving to Christmas—meaning no one in store operations gets to take any vacation days over that time. The nature of the business just doesn't allow for it—seasonal sales increases are huge and even require the hiring of significant short-term seasonal labor. This new head of HR immediately built a strong bond between HR and store operations by having HR "go black" over this period of time as well—even though it wasn't necessary for the daily needs of the business. That particular move let store operations know they had a committed partner—the action drove that understanding better than any words could.

Build Common Goals

Work first on the ideas where you can build a consensus. I have championed many different ideas in my career. The ones that were easiest to accomplish were those that I was able to get others to champion. The reality is that you will never have time to champion *every* good idea you have. Even the most successful and dynamic organizations do not have enough staff or money to work on everything they choose to. They constantly have to prioritize. And prioritizing is not perfect—there is always a certain "fog of war" around it.

So you can always *choose* to work on ideas someone else wants to work on. Champion someone else's idea or agenda. You, as a manager or leader, have the opportunity to invest your time, your "weight," behind those ideas you think make sense for the organization. Find the ideas that someone else has that align with your priorities. If you do, you will have inherent support.

When you come into a new position or organization, take inventory of the ideas around you. Ask questions. Learn what people are working on—and what they wish they were really working on. Find the ideas other people want to work on that match your vision of where you want the organization to go. You will find plenty to work on—and it will all match where you want to take the organization! You will start with people willing to commit resources to your vision, because it is their vision as well.

In Summary

Choose to champion others' ideas.
Support ideas that build toward your vision.
Give others the opportunity to make decisions.

Chapter 9

Create Linkages in Your Management Team

In frontier ground, I would keep my forces closely linked.

—Sun Tzu

NAPOLEON SAID, "One bad general is preferable to two good ones." Why? Because two good leaders create different visions, ideas, and directions that people try to execute. Conflicting direction creates confusion, which never helps in executing results—no matter how good each direction may be. Do you know any great companies with co-CEOs? Schwab, Kraft, and SAP have tried this. These arrangements never last and rarely produce good results. Kraft disbanded this setup and demoted one of its co-CEOs after just two years, announcing declines in sales. Do you know of a great football team that rotates quarterbacks within a game? Almost every year some major college football team decides they have equally talented quarterbacks and rotates them. However, by and large, these are not teams you see in the big bowl games at the end of the year.

Rotating quarterbacks fails because the receivers and the quarterback never develop intuitive linkage with each other—they don't anticipate each other's moves as experienced quarterback/receiver combinations do. The same holds true with co-CEO arrangements. The organization doesn't know whose vision they are supposed to follow. When the CEOs disagree, who does the organization listen to? It creates a situation ripe for dissension and turf wars.

Many companies use a balanced-scorecard approach to drive some level of linkage and success in a variety of important areas. A balanced scorecard measures multiple factors of an organization's success together. Frequently, a balanced scorecard pulls together some combination of financial, customer satisfaction, and employee satisfaction measures. The thinking behind a balanced scorecard is that such diverse measures are frequently in conflict. Lower costs can yield higher profits in the short term, but cost-reduction moves can sometimes reduce customer satisfaction. A balanced scorecard tries to encourage a balancing of these diverse measures. In effect, when practiced at higher levels of an organization, this also balances short-term and long-term performance, as short-term financial performance can sometimes be "bought" at the expense of long-term customer or employee satisfaction.

In buying functions, many retailers use a balanced scorecard to evaluate their vendors. A balanced scorecard ensures vendors are being evaluated on their total effectiveness with the organization. Issues that don't normally factor into a buyer's decision can be taken into account. As an example, on-time truck arrival at a warehouse isn't normally a factor buying organizations take strongly into account. But backups at a warehouse increase labor costs, reduce service levels, and reduce sales. In the normal course of business, a buyer might not see the extra costs at the warehouse—buyers have a tendency to focus on the delivered price of the product. The balanced scorecard encourages taking the warehouse's service needs into account. Thus, buying functions and warehouse functions become linked.

In manufacturing, a balanced scorecard can take into account the different standards of a well-run plant. Costs are an obvious factor. But what about quality? Employee safety? Consumer safety? A balanced scorecard of plant performance keeps perspective on performance against a broad number of measures. No one measure is allowed to dominate at the expense of the others.

Build Common Knowledge

Every organization has its own methods and processes. Most of these are not written down. This common knowledge of an organization is important to daily execution. It is one reason why high-turnover organizations never quite become efficient. Shared common knowledge is the informal glue that helps experienced teams perform expertly.

Webvan was a company that built too much too fast. Webvan was an Internet retailer of the late 1990s that promised fast delivery of groceries, DVDs, and other frequently purchased items. Their business model had many problems—they made "promises" to their customers that were difficult to keep, such as thirty-minute delivery windows in crowded urban environments. But a core problem was that Webvan built out before they knew how to operate and execute their business. So, instead of operating one large money-losing warehouse and delivery system, they built several locations too quickly and ended up operating eight different money-losing warehouses and delivery systems. The business eventually folded.

During the Internet bubble, first-mover advantage—being the first to bring a new idea to market—had a lot of credibility. The thinking was that the first company to bring out a new idea would quickly capture most of the available demand, and subsequent entrants would have to take the business away from the "first mover." When the Internet bubble burst, we were reminded that businesses have to have a business plan that eventually makes money. Building common knowledge is expensive. Building an Internet-based business today capitalizes on the knowledge gained, and mistakes made, by others. Business processes that had been unique are not common knowledge in the industry.

Many Internet businesses now outsource their direct-marketing systems—they let others build the customer relationship management systems they use. With rich customer profiles, these databases are a key asset of many Internet-based businesses. But it is the knowledge owned *behind* the system that builds uniqueness, not the system itself. The longer a company operates that system, the more knowledge that company builds and owns. Putting that knowledge to work, in alignment with the core business proposition of the company, builds the value of the enterprise.

Building on someone else's knowledge is cheaper than building your own. Where possible, consider building on lessons that cost *someone else* the money to learn.

Pick your learning opportunities—and tie them closely to your business strategy. You can learn to do some things well. And that can competitively insulate your company. Dell has become a world leader in just-in-time production of computers. Their latest plant in North Carolina is supposed to be state-of-the art in many ways. There is also a lot of stuff you don't see Dell doing. Their marketing is solid but certainly not spectacular. They ship through UPS. But they build the computers themselves.

Proactively make "buy," not "build," decisions. "Build" decisions have a strong tendency to be decisions where you are creating some new knowledge. If you buy, you are capitalizing on someone else's learnings. This obviously relates to new technology. But it also applies to something as simple as servicing your own trucks or outsourcing that service to someone that only does that. Many high-tech companies buy their customer service call center software off the shelf. That service aspect of technology has been well developed by plenty of call centers that operated for years before us. It is part of that industry's common knowledge.

When making buy-versus-build decisions with technology, consider if you are a technology company or not. If you aren't, then your competitive advantage likely doesn't lie in the technology itself—it likely lies in how you apply the technology. Carefully consider undertaking the cost, and risk, of building significant technology applications yourself. That money may be better invested in operating that technology well and building it into your business operation in unique ways.

Decide to be a slow follower. Innovation feels good, right, and strategic. It is what we are "supposed" to do. But in some aspects of your business, it is okay to be a slow follower. It is how you keep your costs under control.

Many companies do this by limiting the investment of resources outside of the areas "core" to their business. Vitamin retailers like GNC are up on all the current trends in vitamins and supplements. Want to see what's new? Look at the displays at a GNC. They invest in the human talent that is knowledgeable in these areas, attend the appropriate trade shows, and subscribe to the periodicals to know what is happening next.

By contrast, visit your local mass merchant and see what vitamins and supplements are stocked there. They won't carry most of the newest ones. Instead, they will probably occasionally visit a GNC-like retailer, see what is new and what seems to be selling, talk to vendors, and try to add just the best sellers to their own selections.

Common knowledge is an important commodity within the operation of any company. In a younger company, building common knowledge is particularly important—the corporate understanding of how things work together is simply less mature. Young companies will be building common knowledge as they manage through their normal annual business cycles. Younger companies therefore have greater opportunities to get more efficient year to year. You can capture this efficiency by:

Reducing turnover. By reducing turnover, the same people see the problems and opportunity year to year. They will work to get better each time, improving the errors they remember from the prior year.

Documenting opportunities for immediate improvement. In food retailing, you traditionally write down your notes on Thanksgiving the week after Thanksgiving, while the insights and successes are fresh in your mind.

Building simple standardized processes. A simple common form, a "job aid," can help maintain institutional learnings.

In Summary

Build linkage with balanced scorecard reporting.
Decide where to lead and where to follow.
Experienced teams will have the advantage of informal alignment.

孫子

Chapter 10

Use Adversity to Unite Your Team

CREATE ALIGNMENT

The deeper you penetrate into hostile territory, the greater will be the solidarity of your troops, and, thus, the defenders cannot overcome you.

—Sun Tzu

CHALLENGING SITUATIONS give you the opportunity to build new strengths. Sun Tzu writes about "solidarity." Groups of people who have been through difficult times together can develop that solidarity, or feeling of mutual trust. That psychology, the result of having borne difficult situations together, can linger from these experiences. In Sun Tzu's world, that solidarity was generated by overcoming military enemies. In business, these challenges can be the result of competition in the marketplace, shifts in your industry, or internal shifts in your organization—such as a change in management or staff.

An example in the business world is in the growing online DVD rental business. This is a segment Netflix pioneered. For years, it essentially had market share domination in this business. When Blockbuster decided to enter this business, Netflix responded aggressively. Blockbuster is the number one and dominant national brick-and-mortar store renter of DVDs. Blockbuster spent significantly on advertising and promotions to gain share in the direct-to-consumer DVD business. Netflix responded with aggressive promotions and marketing of its own. These expenditures mounted, but Netflix continued to invest. Said Reed Hastings, chief executive of Netflix, "Online rental is the only thing we do, and (our) advantage is focus and desperation. So we have nowhere to go, right? It was win or die, and that's very focusing." And that's an example of how diversity can build strong belief in a particular objective. Blockbuster had thousands of retail stores and a chance to get into a new business—online DVD retailing. Netflix had only the business Blockbuster wanted to get into. Very focusing for Netflix.

Sometimes the Best You Can Do Is All That You Can Do

During the 2005 Christmas season, the New York mass transit system went on strike. It was its first strike in twenty-five years. And the 5,000 buses of the New York Metropolitan Transit Authority (MTA) stopped, and the 230 miles of subway track lay idle. Seven million people had to find a way to work and a way home.

For online grocery retailer FreshDirect, the MTA was the lifeline for employees to get to work. Eight hundred employees a day rode the subway. With a mass transit strike, other provisions needed to be made, or the plant was going to have to close for the duration of the strike. FreshDirect's new lifeline became four rented motor coaches and two vans. These were dispatched on three pick-up runs to major employee population centers—to try and get most employees to work.

You can imagine the relief when the first bus pickup from the Bronx returned with forty people on it. The next few buses were similarly full. And the production facility operated each night during this strike with these transportation arrangements.

Customers were amazed that FreshDirect delivery trucks made it through the traffic and congestion of a transit strike. One e-mail from a grateful customer ended with "Kudos to FreshDirect."

And for FreshDirect employees, doing business in these difficult conditions was a source of pride and accomplishment. It was also a unifying event, because all managers shared the risk and uncertainty of not knowing whether employees would be able to find a way to work and if there would be enough employees to operate a department.

Lessons from CEOs

Businesses frequently use adverse situations to build and unify a team. Xerox is a company that has bounced back from some very difficult times. So difficult, in fact, that the *Wall Street Journal* called Xerox's financial shape in the early 2000s a "near death experience." These difficult times were particularly challenging for the employees of Xerox who grew up in a company that for years held patent protections on key aspects of their business and had traditionally been a benevolent company with a high level of employee security.

Newly promoted CEO Ann Mulcahy used Xerox's challenging financial condition to build commitment to certain objectives and unify her team. Ann talks about having "town meetings" in her early days as CEO, where she could talk about where the company was going and build some confidence that it was going to get through its difficult times. And working with the senior management team in 2000, she says, "We wrote a story that described how various constituencies would talk about us—investors, analysts, reporters—and we dated it December 2005. We took it from the perspective of what the customers would see in terms of what had happened to the product portfolio, keeping it very tangible about what was possible." And she says much of what they put down as their December 2005 goals happened.

Using difficult experiences like this to unify a team and build commitment to a new goal is a common theme among CEOs who have managed through adversity. After Coke had a series of management turnovers and failed products, they tapped a retired Coke veteran to return as the CEO, Neville Isdell. Isdell refused public comment on the company during his first 100 days in this role. He says about this time, "We were working with the top 150 [executives] in terms of building the strategy of what we wanted to do going forward. . . . It probably could have been written in six to eight weeks by six people, but then it wouldn't have been something that people were able to sign on to easily because it would have been what I call the edict from the mount." Leaders use adversity to build a vision, or a plan, that is shared by their key leaders. That vision or plan does not come from the most senior person—like the CEO in the cases of Xerox and Coke. That shared vision comes from making the effort to get other people to contribute to, and buy into, a plan. The adversity these companies were going through added a sense of urgency in building a plan.

In listening to CEOs who have managed through difficult times, you hear a couple of consistent themes:

Seek support. It may be tempting in difficult times to issue mandates. But business leaders who have successfully come through challenging times consistently talk about building support and buy-in for their initiatives.

Create a picture of the future. This gives people something to work for. And something that also lasts when the unifying pressure of outside adversity passes.

Work with all stakeholders. Work with employees, customers, and perhaps vendors and shareholders. Build support for your vision among key stakeholders.

Adversity then becomes an accelerator that lets change occur faster than it would normally occur in less challenging times.

Adversity makes you focus on solving the largest problems. Quickly. While we never seek out this level of adversity—like the Xerox corporate "near death experience"—these experiences can be used to solidify commitment to a course of action that in more peaceful, calmer times would seem less necessary. So adversity—when used with skill—can be used to accelerate a company's progress toward needed and long-lasting change.

In Summary

Use competitive actions to unify your company.
Involve key team members in determining responses.
Use adversity to create needed change.

People Always

Introduction

Mission First, People Always

THE MARINE CORPS uses the phrase "Mission First, Marines Always." It means that while the primary objective is to accomplish the mission, you are going to be able to do so only if you take care of, and utilize, the people you have. Every investment banker, when he talks about the latest company he has invested in, will extol the virtues of its "great management team." Almost any CEO will tell you about the quality of his or her employees. Building great companies depends on great people.

Attracting and retaining great people requires some combination of superior performance in five areas:

Compensation

While salary and benefits are never enough in and of themselves to create loyalty, compensation that is below industry levels or doesn't reward and incentivize performance can cause you to lose good people your organization can't afford to lose. People have a tendency to find the pay level that they can make. At its most simplistic, if you pay $7 per hour, you will have a tendency to attract $7 per hour people. In general, people rise or fall to a certain level of compensation in terms of performance, morale, and initiative. People have a level of pay that they think they are worth. Maybe it is what they have been paid before, or maybe it is what they think they can make, but if you pay more for a given job, you will tend to attract people who think they are worth more.

Training

Good people have a tendency to want to get better. They want to learn new skills. Companies that don't have training programs may lose good people who want to progress in their industry. Conversely, they will keep those employees who are not motivated. That's a recipe for disaster. Unless your organization continues to train and challenge its people, it will experience a "brain drain" of its best and brightest employees. If your employees decide to change jobs within the same industry, your competitor is going to reap the rewards in terms of inside information about your plans and practices.

Perceived Opportunity

People want an opportunity to advance professionally and financially and to contribute a little bit of themselves to the jobs they hold. Employees will stay where they see an opportunity to acquire credibility or a reputation that can serve them later in their career. For years, Procter & Gamble has acquired a stable of superior talent in its brand-management ranks, in part because each candidate understood the occasionally used line "If you get into P&G, you can go wherever you want next." Being "selected" by P&G traditionally has given brand-management job candidates credibility with other marketing-oriented companies. A few years at P&G often allows candidates to write their own ticket at other companies when and if they decide to move on. It also benefited Procter & Gamble—by giving people "portable equity," they had more leverage than their competitors to recruit the most talented people.

Treatment

Caring about people is a significant factor in employee retention. If you can do it with sincerity and some measure of consistency, it can even become part of your company's culture—something that employees feel throughout the organization. Research shows that employees' opinion of an organization is largely based on how they feel about their immediate supervisor. Thus, treating people well needs to become an expected behavior at all levels of management.

Values

For years, the Container Store has been named one of the top companies to work for in America. They give a lot of the credit for this achievement to their adherence to their six foundation principles. Simple principles about what they believe about how they do their business—how they treat customers and their own people. As an example, one of the Container Store's foundation principles is, "Fill the other guy's basket to the brim. Making money then becomes an easy proposition." The "other guy" is employees, and customers. In other words, you make money by growing your base of loyal, satisfied customers—and having satisfied employees who can deliver the level of service that makes customers loyal. If you have a lot of satisfied customers, you can build your sales and profits—financial success follows the customer-satisfaction success. That makes sense—and that also sounds like a company many people would want to work for. If they live values like those—if those values are transferred down through the management ranks—that would be considered a special company by many current or prospective employees because principles like this put an emphasis on how you treat others. That's the kind of value that resonates with employees.

Companies that have high employee turnover have great difficulty executing results and great difficulty succeeding. Companies that retain, train, and motivate their employees better than their competitors win at execution.

Chapter 11

Hire Skilled Commanders

[A skilled commander] is able to select the right men and exploits the situation. He who takes advantage of the situation uses his men in fighting as rolling logs or rocks. It is the nature of logs and rocks to stay stationary on the flat ground and to roll forward on a slope.

—Sun Tzu

THROUGHOUT MILITARY AND business history, small outnumbered armies have succeeded against much bigger foes. The American Civil War is full of such stories. During the early years of the war, the Southern generals—Robert E. Lee, Stonewall Jackson, Jeb Stuart—won victories against the more numerous and better-equipped Northern troops. They won because of their greater skills—daring tactics, skilled maneuvering, and a sense of when to take a risk.

Skilled commanders, and managers, seem to be few and far between. Why is this so? Businesses are very complex, requiring the application of a large set of skills. One military historian writes "in any organization the opportunities for misplacement of personnel (or anything else) go up as the square of the complexity." The square of complexity simply means as we add complexity to any situation, the likelihood of error goes up. And business has gotten more complex. Managing a business of most any size now requires management of complex IT decisions, negotiating increased government regulation, and making changes with great speed. Fitting the right people into the right jobs is more challenging because the *content* of many business challenges is growing. Managing in large companies, in particular, requires a lot of judgment on what areas of the company to involve in solving a problem. Many issues can require an assessment of PR, legal, or employee (union?) implications.

Also, the strength of the status quo frequently outweighs the need to take measured risks. Wall Street values predictability of results. They do not like surprises in how a company performs. Success is measured by meeting Wall Street's prescribed benchmark—whether that benchmark is realistic and informed or not. Accordingly, in many industries, privately held companies outperform publicly held companies. Privately held companies can choose to take risks that might cause a few poorer quarters but result in the long-term health of their business. Publicly held companies have great difficulty doing these things—they must make each quarter's results.

Or politically able leaders fall into roles they are less qualified for. As one commentator said, "Quite naturally, the least gifted commanders are usually the most political."

Develop Your Leaders

The greater complexity of business decisions—and the greater breadth of knowledge required of business leaders—makes developing employees increasingly important. We count on key people not just to solve problems, but to identify when to bring in additional resources. Managers need to assess and fix problems and know when they need help. In part because competitive situations can change very quickly, and in part because businesses are under greater scrutiny.

One example of this greater scrutiny is the media. Media outlets covering business have multiplied—we have business-only cable stations and radio stations and Internet sites covering many specific business topics. As media has segmented to serve different customer groups, the absolute amount of business reporting has greatly increased. As new media have grown up, new forms of reporting on businesses have come with these new media. Satellite radio spawned new business programming. The Internet allowed the growth of blogs—Web logs written by individuals. Blogs can increase dramatically the amount that is written about a company, an individual, or a specific issue. With radio, TV, and the Internet, much writing and reporting happens in real time. So business responses have to be quicker.

Business managers need to apply a lot of judgments to their work. They need to bring business experience, and add to that an ability to handle new situations that are outside of their direct business experience. For business leaders with the right training, these new challenges can be exciting and rewarding. Without the right training, companies, annual budgets, or promising careers can be put at risk.

For many managers, developing people who can handle new challenges is one of the highlights of their jobs. They may work for companies that put a premium on training and growing employees. Or they may value employee development based on some experiences from their own working careers. But either way, there is probably some combination of the following techniques that they use to develop employees.

Stretch Jobs

Research shows that peak learning experiences happen in jobs where people are challenged to work outside their comfort level and learn new things. During these times, employees will invariably experience stress. They are working extra hours because they are in unfamiliar territory. Their morale may be lower than normal because they are not sure they can handle the extra responsibilities. They may fear failing.

When taken to an extreme, or for too long, this is obviously counterproductive. But the right amount of stretch creates a positive stress—the stress of successfully negotiating your way through some unfamiliar territory. If you give your leaders enough room to succeed, and support them through these experiences, these can be great ways for new leaders to develop. It can also be a tremendous bonding experience for your organization.

New Roles

Some companies proactively switch managers between roles. This can happen as part of a learning experience—some companies recruit new managers into programs where they get to work in many different parts of the organization as part of their training and orientation. Many retailers, when they hire people for their headquarters organizations, first put new hires to work for several months in store operations. They do this because companies find this is a great way for new employees to learn the guts of their retailing operation. If you are a stable organization with seasoned employees, allowing people to switch roles temporarily can be a low-risk way to teach people new things—and to improve communications within your organization. You will also be cultivating employees who have a more holistic view of your organization and bring out-of-the-box thinking to different areas of your organization.

For your own career, look for jobs that give you some new experience. Many good careers involve some lateral moves that provide new learnings. Don't always look for the next promotion—promotions likely take you further up the same functional silo you are already in. In addition to promotions, look for jobs that will give you a chance to add a new skill to your background.

If you are in marketing, consider a job in sales. If you are in IT, consider a job in operations. Any new skills you learn are your portable equity. The more skills you have, the more choices you will have in the future. And the more valuable you will be to your employer.

Take these skill-building career moves yourself—and help your employees into positions where they get to add to their skills as well.

Mentoring

Most of us have learned our craft from a few select people. It's not that we haven't worked with a lot of different people, read to keep abreast of our industry, and attended different seminars. But most people learn most of their knowledge from a relatively small number of people in their career. Sometimes we find a boss or a colleague who is willing to teach or coach us—those people are invaluable.

Training and mentoring experiences can provide needed learning. Feedback and insight are best delivered at the time employees can use it—don't train in advance of the need. Training and mentoring are most effective when done by fellow employees or experienced teachers who understand the organization and can make the training immediately relevant.

If you are looking to develop yourself, take advantage of opportunities like these. Make a list—which of these are available to you? How could you create opportunities, or relationships, that would increase your own development? As you grow your company's business, don't forget to invest in your own skills development. Remember some skills are technical, like writing a sales plan or running a particular piece of equipment. Other skills are more broadly applicable, like being able to speak in front of a group. Over time, try to develop some of both.

Make the Most of Your Team

Don't ever assume you will have exactly the team you want. Any team you manage will always have some shortcomings. A manager I once worked with would say "Never trade the eight of diamonds for the nine of clubs." He meant that you fire someone only if you believe that you are going to be able to get a much stronger performer in that person's place. It is not a win if you lose someone and trade up only a little bit. The costs of firing someone, having a job vacant, and having to train a new hire are not inexpensive. Constantly replacing people in an attempt to end up with the "perfect" team is ineffective. That can lead to high turnover, and people you want to keep may come to distrust their future in the organization.

The skillful manager gets the right people in the key positions and makes all of the people around him or her better through setting clear objectives and providing training and development. By doing this, the team builds competence *and* trust. Two assets necessary for executing results.

Prior to World War I, the Prussian army regularly played a practice war game called *Kriegspiel* to train its officers. It was played with wooden blocks on a map. Unlike other war games, the participants could not see what their opponents were doing. Nonplayer umpires oversaw the game and implemented orders from the two sides.

One military historian writes about Kriegspiel, "The victor was usually the player who anticipated the short-comings of both his subordinate units and bad information." Assuming imperfection, and managing skillfully, is a leader's job. It is also one of the most critical aspects of being a skilled commander. It is one reason why experienced leaders frequently outmanage inexperienced ones. Experienced leaders have failed frequently—both on their own and with a team of people. Because of those experiences, those leaders are better able to anticipate where failure is likely to come from, and minimize the effects of that failure.

Great companies are built on great people. And building great people starts with great commanders. When you find them, hold on to them.

In Summary

Help your employees build new skills.
Give your people peak experiences.
Look for peak experiences for yourself.

Chapter 12

Take Care of Your People

If an army encamps close to water and grass with adequate supplies, it will be free from countless diseases and this will spell victory.

—Sun Tzu

PERENNIALLY ONE OF the top 100 companies to work for in America is Wegmans Food Markets. A regional purveyor of premium fresh and prepared foods, Wegmans adds value to its products with the services customers receive at store level. President Danny Wegman teaches the spirit of "caring about one person at a time."

Caring about one person at a time is not as easy as it sounds. It is never convenient—people don't schedule their problems around your convenience. And it requires action—timely action. But it may be the best way to build a culture of caring for people. Stories spread—stories that start from the top.

This is an attitude that is shared among great retailers. Great retailing, brick-and-mortar retailing in particular, is driven by great execution. It doesn't matter how good the products were, if the checkout lines were five people deep. You won't go back. The greatest retailing strategies can easily be defeated by the high school checkout clerk who is the last person you meet at almost any store.

Caring about people helps them work together—and that shows up at the bottom line.

Mike Bingham, who has headed large production operations for companies like Safeway, Rich's, and Del Monte, tells the story of taking over a Frito-Lay production plant that was "broken" in its relationship between managers and employees. They weren't working together effectively. So the plant wasn't working effectively. Mike fixed the plant by learning the names of all 700 hourly employees. That took him four months. And within six months he had the plant's management and employees working together, and plant productivity was back to meeting expectations.

Many successful leaders work hard to have employees feel good about the work they do. Looking employees in the eye and saying words like "That is an important job" and "Thanks for being here with us today" can make people feel the work they are doing is valuable. Employees who feel their work is valuable will stay with a company longer.

Build the Right Work Environment

The work environment isn't just physical—it is the spirit of the company as well. Many companies today employ a large number of first-generation immigrants. Production facilities that have high employee-retention rates generally provide a language-friendly plant environment for these employees. Examples include bilingual supervisors and managers and promoting workers into management ranks—stories about managers who started as hourly workers and worked their way into plant management can be powerful for attracting and keeping other employees wanting to travel that same path.

Building the right work environment can translate into innovative policies too. FreshDirect has found over time that first-generation employees generally go home and visit their families for an extended time period each year. Airfare is not cheap, and their home countries are not close by. Thus, these trips can last several weeks. FreshDirect has always found ways to accommodate this so that the best employees will stay. However, people were taking their leaves at times that were not convenient to the business. FreshDirect's business has seasonal spikes, and losing good people at the wrong time can be painful. Inspiration hit when a chef told a story about working on a cruise line. When an employee completed a "tour" on a cruise ship, went home, and came back, the cruise line would reimburse his or her airfare. However, the airfare would be reimbursed only after the employee came back.

FreshDirect put a similar program in place for hourly employees—"Summer Leave." The Summer Leave program matched up a seasonal trough in business with an incentive to take a long break from work. This incentive is open to any employee who leaves work for four to eight weeks in July or August—the slowest time of the year. Summer Leave applicants have their job guaranteed if they returned to work by mid-September. Upon return and completion of a few weeks worth of work, they receive a check—essentially enough money to pay for airfare home.

To date, more than 95 percent of the people who leave for the Summer Leave program come back and claim their extra paycheck. A win for employees, and help for balancing a workforce across the year.

In Summary

Your people represent your company.
Provide the right working environment.
Build employee programs that help you structure a profitable business.

Chapter 13

Share Rewards

Now, in order to kill the enemy, our men must be roused to anger; to gain the enemy's property, our men must be rewarded with war trophies.

—Sun Tzu

MOST BUSINESS LEADERS accept the thought that great companies are built by great employees. A company's creative ideas, and winning execution, come from the work of its employees. Retention of great employees is determined by how you treat them, how you train them, and how you pay them.

Good people have a tendency to know their value and what they contribute to an organization. So if you want to retain good people, you need to pay them what they are worth. If your work environment is demanding or the hours are unusual, you will need to compensate accordingly.

Companies that try to keep good people without investing in them often end up being the loser in the war for talent. An employee who doesn't get training but wants to better himself will seek out an employer that will develop his skills. Employees who know they are doing a good job will want to be paid accordingly for that work.

Jim Senegal of Costco is repeatedly asked questions about why he pays his people so well. Jim says simply, "We pay much better than Wal-Mart. That's not altruism. That's good business."

Costco's CFO expounds on this thought more specifically: "From day one we've run the company with the philosophy that if we pay better than average, provide a salary people can live on, have a positive environment and good benefits, we'll be able to hire better people, and they'll stay longer and be more efficient." Senegal's strategy of taking care of his employees above and beyond the call of duty has not gone without criticism. One analyst famously said of Costco, "It's better to be an employee or a customer than a shareholder. He's right that a happy employee is a productive long-term employee, but he could force employees to pick up a little more of the burden."

However, Costco's stock continues to do very well, trading at a price/earnings multiple higher than most of its competitors'. Costco's head of HR says, "When Jim talks to us about setting wages and benefits, he doesn't want us to be better than everyone else, he wants us to be demonstrably better." Senegal has unapologetically stuck to his guns, saying in one interview, "We want to build a company that will still be here in 50 and 60 years." And he believes maintaining a highly stable workforce is part of how he ensures that future.

Successfully retaining people requires many things:

Rewards that are relevant to the employees. Different employees want different rewards. An older workforce will place more value on good health-care benefits than a younger, single workforce will. Your reward structure should conform to the needs of your workforce, or the workforce you are trying to attract.

Consistently communicating the value of what you offer. Beyond a weekly paycheck, benefits or rewards you offer will not tend to register with your employees. A pay raise (or bonus) is easy to understand—employees see that in their weekly paycheck. Paying more of an employee's health-care benefits, for example, is more difficult for an employee to understand the value of. It's not money they can spend as they wish. Thus, most employees don't understand the significant expense it entails for your company—and the commitment to the employees that it represents.

The only way for your employee to understand the reward you offer is to communicate it to them—directly and consistently. You need your employees to understand the value *they* get from the money you are spending. If not, these are not good expenditures—or good rewards.

Managers and Supervisors must understand the rewards. Can you, your managers, and even senior managers recount the benefits offered by the company? The fact is that most managers have gaps in their understanding of their company's benefits and nonpay rewards. Many employees will talk with their direct supervisors regarding their company's benefits and nonwage pay policies. Make sure your managers and supervisors have adequate knowledge to answer those questions accurately, on the spot.

Your employees need to understand what they are receiving—and how it reflects the company's commitment to their talent. Constantly communicate the value of your benefits to your employees. This is particularly important with nonpaycheck rewards such as flextime, casual dress policies, and so forth. If your employees don't know how what you are offering differentiates you from the competitors that want to hire them, they can't find value in that reward. The details are important here. Do you have a diverse staff? How many languages do you need to communicate in? Are certain benefits more valuable to certain employees (e.g., flextime for working parents)? Don't rely on one single communication—too many people will fail to understand what you are trying to say.

A former regional manager of McDonald's tells how he retained employees with attendance contests. The McDonald's workforce is young and highly mobile. A minimum-wage job at McDonald's isn't that different, or better, than a minimum-wage job somewhere else. So one strategy to retain young workers was contests. When the store reached a certain goal, the entire store won a bus trip to the local water park (a crew from a neighboring store would be brought in to run that location for the day). For that particular workforce, that was a relevant reward—and one that differentiated that employer from other minimum-wage choices.

In Summary

Pay what you need to pay.
Rewards should conform to your workforce.
Communicate to employees as effectively as you do to your customers.

Chapter 14

Motivate

Hence, the general who advances without coveting fame and retreats without fearing disgrace, whose only purpose is to protect his people and promote the best interests of his sovereign, is the precious jewel of the state.

—Sun Tzu

PEOPLE ALWAYS

MANY DIFFERENT COMPANIES in the retail and restaurant industry employ "secret shopper" programs. Employees of a third-party company pretend to be average customers so they can "shop" the store incognito and observe how the store is run, how employees interact with each other, and the level of service customers receive. Some retailers use the program to reward quality work from their employees, whereas other companies use these programs to identify undesirable behavior. Time after time, David Rich of ICC Decision Services finds that the companies who use these programs to *reward* good behavior make more money than those companies who use them to catch employees unawares. Reward, rather than fear of being chastised, is a bigger motivator for employees.

In secret shopping as in life, a positive attitude wins over a negative one. To build effectiveness, leaders need to inspire. They need to get the most out of those around them.

How you set up a situation will determine how your people will react. The French military commander the Marquis de Saxe explains it the following way: "Thus when you have stationed your troops behind a parapet, they hope, by their fire, to prevent the enemy from passing the ditch and mounting it. If this happens, in spite of the fire, they give themselves up for lost, lose their heads, and fly. It would be much better to post a single rank there, armed with pikes, whose business will be to push the assailant back as fast as they attempt to mount. And certainly they will execute this duty because it is what they expect and will be prepared for." Don't set up people into a situation they are likely to fail. Their reactions when faced with failure may not be what you want. Set up situations with positive, realistic expectations.

Sun Tzu expresses this aspect of motivating people as well, similar to the Marquis de Saxe's experiences: *Now, at the beginning of a campaign, the spirit of soldiers is keen; after a certain period of time, it declines; and in the later stage, it may be dwindled to naught.* The Marquis de Saxe did his writings in the 1700s while Sun Tzu wrote in 500 B.C. However, the underlying principles of getting the most out of a group of people don't change. How you build the expectations of your team will determine how they react. And that will affect their motivation, and your ability to retain their commitment to your goals.

One organization that is a great example of maintaining the motivation of its employees is the New Seasons market in Oregon. At this small retail chain, employees are given "get out of jail free" cards and permission to do what the customer wants. Printed on the back of the card it says:

Dear Supervisor: The holder of this card was, in their best judgment, doing whatever was necessary to make a happy customer. If you think they have gone overboard, please take the following steps:

1. Thank them for giving great customer service.
2. Listen to the story about the events.
3. Offer feedback on how they might do it differently next time.
4. Thank them for giving great customer service.

This system gives employees permission to take a risk, and supervisors an open and comfortable way to correct behavior and provide training.

One of the roles of a business leader is to get other organizational leaders behind you. These may be peers or subordinates. Getting peers and subordinates to commit to your direction is a great step. As you do this, make sure you are leading people down a path they can succeed on. As a leader, your role is to clear obstacles in their way. Anticipate where internal opposition may come from, and work to manage that opposition. Successful leaders:

1. Know who their allies are and win as many senior-level allies as possible to their "cause" early,
2. Know what objections they will hear, and set up early tests to assuage those objections, and
3. Find a way to attach their goals to other goals the company is already pursuing and committing resources to.

Build High-Quality Management

Research shows that employees will judge a company based on their opinion of their direct supervisor. That relationship defines your company's "brand" to your employees. Thus, your organization needs to make its managers and supervisors as good as they can be. Hiring practices and training practices will dictate how this is done for your particular organization.

My father, business author Gerald Michaelson, used to say, "It's not the people you fire that cause you stress, it's the ones you don't." In all these decisions, balancing your responsibility to the company, the individual manager, and the people underneath that manager is challenging. How you make these decisions is important—these decisions impact the culture and the trust employees have in the company. And they impact the trust people have in you. Create trust by:

Maintaining high performance standards. During the course of my management career, I have been hired by several poorly performing organizations to improve their processes and performance. In these situations, you will often find people that by objective standards are not performing well. And frequently "the bar" has been allowed to slip. Performance that is not acceptable had been judged acceptable—otherwise the poorly performing employees would have left on their own accord when they realized they didn't pass muster, or they would have been replaced. I start in these situations by setting—in effect "re-setting"—the performance standard. It could be a process, a measurable result, or a style of interacting. Reset these standards with firmness and also with empathy—failing to meet them immediately doesn't mean the employee will be fired, but he or she will be made aware if he or she doesn't meet the standard. Frequently, you will find many employees, with a little help, rise to the new expectations. People want to succeed.

Use metrics. Emotion is sometimes a useful business tool—but more often than not, it isn't. Most emotional conversations lose some logic and rationality—and can harm business relationships. Setting some simple metrics gives a "grounding" to conversations and keeps things focused on business performance. For example, don't allow your employees to rant and rave about how a particular employee or policy upsets them. Instead, have them focus on how it impacts their ability to do their job. Is the problem at hand diminishing their productivity? Costing the company customers or profits? What are the metrics of the situation? Setting the right metrics lets you focus your conversations on the most leverageable points for your business.

Always respect the individual. Why? You will be more effective. People listening to you will actually hear you better, and they will take your feedback less defensively. They will be more likely to change their performance based on your feedback.

If this approach sounds too soft, don't forget people are watching you as a manager. How you handle these situations will define how people think of you. And these situations will significantly impact how your people believe they will be treated if they were placed in a similar situation with you. Put in place practices that support your motivating, your getting the most out of, the people that you have.

In Summary

Define expectations.
Make the decisions that need to be made.
Be fair to individuals.

Chapter 15

Execution Is Driven by People

When the officers are valiant and the soldiers ineffective, the army will fall.

—Sun Tzu

START WITH THE best people you can hire. The vice president of operations for the Container Store, one of the best companies to work for in America, according to *Fortune* magazine, says, "One of our foundation principles is that one great person equals three good people. If one great person equals three good people, one good person equals three average people, and one average person equals three lousy people." With that math, it is clear why the Container Store is so adamant about taking their time in hiring people. Experienced companies offer some thoughts on what to look for in great people:

Look for "unteachable" strengths. In football, there is an expression: "You can't coach speed"—either you have it or you don't. In business, the same principle holds true. Depending on the job, a certain level of creativity might be needed. In sales jobs, an outgoing personality may be a requirement. These are traits that not all people possess and can't be taught. You just have to hire for them.

Look for a winning track record. Winners have an appropriate track record. People who have built a business, successfully introduced a new product, or fixed problems know what that success feels like. They likely have the confidence to overcome obstacles.

Look for fire in the belly. Call it character. Call it desire. Jimmy the Greek might call this "the intangibles." Someone who will have the determination or resilience to make the right things happen. At Procter & Gamble, "fire in the belly" is one of the traits they recruit to.

Look for passion. Someone on a mission can accomplish change where others would struggle. This is one of the characteristics top employer Yahoo looks for in its employees. Says one Yahoo senior vice president, "We want people who are passionate about their subject areas." They even put it on their Web site: "We are looking for people who believe passionately in our mission."

Though talent and track record are critical, you also want to hire people who have made a mistake or two along the way. Perhaps a bad career move? Perhaps a business decision that turned out to be wrong? Someone who has made a mistake—and can fess up to it—has gained maturity. They may also have gained some learning on someone else's dime, so they're less likely to blunder in *your* organization.

Make Plans Your People Can Execute

When the U.S. space program put a man on the moon, our spaceships were relatively simple in overall design. Essentially they were rockets with a capsule on the top. If you look at the earliest footage of rockets, you can see the clear lineage to the Saturn V rockets of the Apollo program. Over time, that program was replaced with the more complicated design of the space shuttle. Designed for a tougher list of requirements—and no longer fully disposable—the main capsule, now a plane, was reusable. The payload was increased significantly, and technology was added to accomplish more tasks in space. This complexity provided a variety of problems: temperature-sensitive O-rings, falling insulation, and damaged tiles. The space shuttle had two massive failures and rarely flew on time after those problems.

NASA has announced that in a few years, the shuttle is slated to be replaced with a much simpler design, likely a rocket with a capsule of some design on top, in concept very similar to the old Apollo design. Simplicity wins out over complexity. Why? Because it is easier for fallible humans to execute. That leads to better results.

Make your plans realistic for the people involved. Set stretch goals, of course. But fit plans into the skills and experiences of the people who will carry them out. As you do this, make the ideas you work on the ideas your people (including peers and others) are excited to execute.

As you work to influence people, you need to be skillful in setting up how people view the ideas you expect them to contribute to. Consider:

Timing. An operation I managed had to cancel almost two days of production at the height of our busy season due to a huge snowstorm. People worked hard getting the production done, they worked hard getting to and from work, and they were tired. But we needed to crank up production to get out more orders. I waited a day to ask about raising production later in the week—it was the soonest we could take more orders—and I didn't lose anything. The one day of rest helped calm frayed nerves, and people willingly agreed to take more orders. They knew it was what the company needed to do, and they were now in a mindset to make it their idea. As you would expect, they executed the incremental production very well.

Championing someone else's idea. If you know what your objective is, you can use more than one path to get there. Let someone else choose the path. You will still get to where you are going. But you will have someone else committed to the objective. Focus simply on moving the business forward—let someone else guide the tactics.

Managing what people are thinking about. Set the tone. Ask the right questions. If you are asking questions, and getting other people to help answer those questions, you are setting an agenda. The organization will concentrate on the things you want them to. Don't micromanage and don't be controlling. Instead, keep working on asking the right questions that will lead to fewer problems, new products, and better opportunities for your organization. Thought will lead to action. You will just have to guide the enthusiasm.

The best leaders make the people around them even better. They have the ability to help their people to be successful. Some do this by setting a very clear direction—everyone knows where they are going and what the organization's mission is. Some do it because they have a great grasp for the business and years of experience under their belt, and they have the ability to share that with those around them. Size up your strengths, and work to get the best out of the people around you.

In Summary

Hire quality.

Look for a winning track record.

Look for leaders who make people around them better.

Flexibility

Introduction

When You Have an Opening, Take It

IN BUSINESS, many companies fear their more successful rivals—those competitors that exceed them in profitability, product design, and market share. However, the competitors to fear are not the ones who try and succeed but rather those that try and fail—and then dust themselves off and try again. Those are the competitors who will, over the long term, excel in their industry and with their customers.

In the early 1980s, Cadillac brought out a small luxury car to compete with BMW and similar competitors. This seemed like a natural offering for the Cadillac brand, which had been associated with luxury but in recent years had acquired a staid perception among consumers. Cadillac's Cimarron offering was an upgrade of a Chevy Cavalier. Unfortunately, the Cimarron was an obvious cousin of the Chevy—not a good thing in the luxury market. The Cimarron flopped. First-year sales were just one-third of Cadillac's expectations, and several years later, the car was discontinued.

In the late 1990s, Cadillac went after this market segment again. Their Catera was again built off of a standard General Motors platform, but this time from a European division. This offering also had a little more differentiation from the core General Motors line. Unfortunately, the car got ho-hum reviews and its "Caddy that zigs" advertising was widely derided. The car sold poorly and was discontinued after four model years.

A few years later, Cadillac tried again with another small luxury car—the CTS. This time their CTS was a sales hit—it was nominated for several Car of the Year awards when it was introduced. It revitalized the Cadillac brand and overall sales for the company. It was a testament to Cadillac's try-and-try-again attitude.

Fast-changing markets illustrate the importance of bouncing back from failures:

- Apple, struggling but never quite completely down for the count in the competitive home computer market, made a significant relaunch with the iPod. Apple's iPods became a cult hit and grew to outsell its computers by more than 10 to 1. Not what you would have necessarily expected from Apple ten years ago when Michael Dell suggested the company should be closed down and any cash returned to shareholders.
- Las Vegas is the hottest hotel market in the country. Steve Wynn's new hotel in Las Vegas, Wynn Las Vegas, at the site of the old Desert Inn, is one of Las Vegas's most popular and talked-about hotels. Undeterred by critics, Steve just keeps on building. He started by buying Las Vegas's downtown Golden Nugget Hotel and renovating it; later, he built a Golden Nugget Hotel in Atlantic City. Then he sold the Atlantic City property and bought a piece of land on the strip in Las Vegas and built the Mirage Hotel. He followed that up with the Treasure Island and the Bellagio. After selling those properties, Steve opened Wynn Las Vegas on the site of the former Desert Inn. By many people's measure, Steve Wynn has never really failed. But he does keep on trying and trying—and successfully beating his own personal best each time.

Manage Your Risk

Trying and failing can be scary. In some organizations, it can even be career threatening. But, if managed well, failing can actually lead you to greater success down the road. The key is to manage failure strategically:

Keep losses small. Testing is part of any business. Testing in small ways lets you learn at a rate the organization can afford. Picking a specific geography, group of customers, or a small set of locations can give an opportunity to learn—without losing everything. Picking a spot that you can monitor closely, and be involved in, can help you maximize your learning. A series of small tests can let you steadily improve a concept and make improvements as you go.

Involve high-level management. Involving top management gives you buy-in and support. It also attaches visibility to your project and can help your test efforts succeed.

Acknowledge the possibility of failure. Expectations can help a team of people support an effort that may not succeed. Do this by building an understanding of the importance of the test itself—how it can yield important information that could possibly build a more successful product or service.

There is an expression from World War II, "The first wave dies on the beach," which refers to the staggering 80 percent loss of the first wave of troops that went ashore on D-Day. If your initiative is that "first wave," it might not work the first time. Manage expectations and build support for the next wave or the follow-up initiative.

Celebrate the win in the loss. This is failing forward. If your test leads to another test, claim victory! Do whatever you can to maintain organizational excitement for the effort.

Manage risk well so that you can tackle bigger risks over time. Successful leaders look to shoulder, and manage, more and more risk. Not unwisely taking risk; rather, in the spirit of creating opportunity, building profits, and growing a business. Managers who can manage risk with success are invaluable to the companies they work for.

Chapter 16

Speed Has Value

Thus, while we have heard of stupid haste in war, we have not yet seen a clever operation that was prolonged. There has never been a case in which a prolonged war has benefited a country.

—Sun Tzu

SPEED PAYS OFF. When Hurricane Katrina hit New Orleans, many governmental and business organizations were caught badly off guard. Initially, Wal-Mart was one of them. However, Wal-Mart quickly recovered and got huge positive PR for its speedy response. Days after the hurricane hit, Wal-Mart chairman Lee Scott was at a press conference with former presidents Bill Clinton and George Bush Sr. He quickly offered all displaced former New Orleans employees an immediate job at any Wal-Mart in the country—and got huge positive press. Almost before FEMA figured out that there were people stranded at the New Orleans Convention Center, Wal-Mart had a public relations coup.

Slowness rarely wins the game. In World War II the German blitzkrieg was moving rapidly across Western Europe—so rapidly in fact that many German commanders were surprised by their own success. Instead of continuing to move quickly against the surrounded British army at the French port of Dunkirk, the German army stopped because commanders were worried about stretching their forces too thin. "Dunkirk was to be had for the asking," writes military historian Kenneth Macksey. And hundreds of thousands of Allied troops lived to fight another day—and eventually defeat German forces.

Moving quickly allows competitive advantage. Auto companies try to speed up their product-development cycle to get new products to market faster. Political candidates try to convert polling data into TV ads and election tactics quickly. Retailers move quickly to sell what is in fashion at the moment.

Moving with deliberate speed is more important in the age of the Internet. Blogs can build a rumor into a reality—and if not addressed quickly, unsubstantiated and false claims can take on an aura of truth. Or blogs can be used to build the mystique of a product.

In today's competitive marketplace, speed is an offensive weapon. Use it:

To reach your market. Regional food retailer FreshDirect was the first to market with a new packaging technology that steamed meals. FreshDirect got incredible coverage in the *Wall Street Journal* and *New York Times*, as well as in regional media. A big coup for a small company! Being first to market with this product created a lot of positive buzz for the company and reached many potential new customers.

To organize decision making. One friend of mine says, "Push decision making to the lowest level possible." That's a scary concept for many companies. What if "they" make a mistake? The thing is that "they" will fix it most times before any senior leader ever learns about the particular mistake. Decisions made by the lowest level are decisions made in real time. The empowered hotel clerk who gives a customer a room upgrade and the customer-service representative who satisfies a complaint are examples of people at the lower rungs of an organization intercepting a problem or an issue quickly. Customers like this, and handling customer complaints with speed and fairness frequently can turn a bad customer service encounter into a loyalty builder.

To gain actionable intelligence quickly. Larry Weiss of the research company Linescale uses online consumer testing to give his clients the advantage of speed. Answers to online questions come back in days, and results are instantly tabulated. This approach to research is much faster than phone or mall research, where results take several weeks to be processed. Faster knowledge can lead to faster action.

Create a Sense of Urgency

A sense of urgency drives results. This is obvious in times of a national emergency such as the 9/11 terrorist attacks, the 2004 Indonesian tsunami, or Hurricane Katrina's devastation of New Orleans. In each case, corporations, government, and private citizens responded to support relief. In the business world, we rarely have such obviously galvanizing events. But there are still many ways to create a sense of urgency:

A competitive threat. You can use the growth and success of a competitor as a call to action for your employees and to build a sense of urgency for creating change in the workplace.

Customer data and stories. If you don't have customers, you won't have a business. Trend data on customer satisfaction, or even anecdotes from customers, can build urgency for creating change in your products or services that the customers will see. Most employees know that customers are important—without them, the employees wouldn't have a job to come to every morning. Use your customers to communicate to your employees—be it directly or indirectly, through surveys and research—it will have more impact than if you say those things yourself.

Seasonality. For many consumer companies, the fourth quarter is a huge quarter—the Christmas holiday drives an obvious increase in sales. You can use that as leverage for getting projects completed "before the big surge." You can also make an appropriate call for extra help and work during a busy time.

Sun Tzu was a big fan of speed—provided it was used opportunistically. He writes, *By "situation," I mean he should act expediently in accordance with what is advantageous in the field and so meet any exigency.* Speed, when exercised with good judgment, is a practice that leads to success.

In Summary

Move quickly when opportunity arises.
Prepare for opportunities to enable speed.
Deliberate preparation enables speedy action.

Chapter 17

Use a Variety of Attacks

In battle, there are not more than two kinds of postures—operation of the extraordinary force and operation of the normal force, but their combinations give rise to an endless series of maneuvers. For these two forces are mutually reproductive. It is like moving in a circle, never coming to an end. Who can exhaust the possibilities of their combinations?

—Sun Tzu

BUSINESS IS rarely predictable. We all make plans. However, those plans must change as the business develops. In the military operations, plans change when contact with the enemy is made. Only an unskilled or foolish commander insists on rigidly sticking to advance plans if a battle requires alternative measures.

Says Clayton Jones, chief executive of defense contractor Rockwell Collins in the January 9, 2006, issue of *Forbes*, "We compete in the land of the giants. What we lack in size we make up for with focus and agility." Jones believes he manages this agility by listening to his customers, who give him new insights as the business battle unfolds.

Netflix invented the online rental of DVDs. In 2005 Blockbuster decided to take its shot at owning that business. Netflix responded by spending aggressively to grow its business, add new subscribers, and maintain its lead in this business. Netflix significantly changed its plans—and watched its stock price plunge from $39 to $9.

However, the strategy proved effective. At the end of that year, Blockbuster crowed about its success at achieving over 1 million subscribers. Netflix, however, remained the dominant industry leader. Its decision to spend aggressively had grown its subscriber base to *over 4 million subscribers.* And by the end of the year, Netflix's stock had rebounded considerably based on its success in growing and maintaining its market dominance.

Netflix changed course and responded well to changes in the marketplace. It deviated from its original plan. A mid-2004 financial release stated, "The Company said it expects to increase profitability through 2006." That clearly didn't happen—though it might have been a reasonable expectation at the time. The situation changes the plan.

Bring a New Perspective

At the World Economic Forum in Davos, Switzerland, consultant Tim Brown advised, "We learn our way to solutions." He advised finding a problem you have never worked on before to allow you to bring a fresh perspective. In other words, he advised, approach problems with a beginner's mind. A new view of a problem is valuable—in meetings and deliberations, strive to get a variety of opinions heard. This simply leads to better decisions. Do this through:

Inviting someone who will likely hold the minority view. Perhaps that is someone from outside your department. Or someone who comes from a different set of experiences (either from outside of the company or someone who came up through a different functional silo within your company).

Encouraging dissenters to speak up. Hear opposing viewpoints. Seek out different opinions. Conduct meetings in a way that makes people feel comfortable and rewarded for speaking up.

Asking open-ended questions. To encourage alternative viewpoints, direct your open-ended questions to people who (1) are likely to disagree with you or (2) have not been major contributors to the conversation.

Research has shown that the IQ of a group is higher than the individual IQ of any one member of the group. This makes sense because the group can use all the talents and experiences of all people in the group. Use this math to your advantage in building your business. But the math adds up only if the contributions of each and every team member are sought. If the team is led by a bully or someone who listens to a chosen few, the IQ of the team may never exceed the IQ of that person.

Sometimes, it is simply a matter of style in hearing new perspectives and allowing you to consider a broader range of choices. An open, collegial style can encourage more open debate and sharing—but only when practiced consistently and with sincerity.

Sun Tzu recognizes this in his management of armies. He advises, *It is the business of a general to be quiet and thus ensure depth in deliberation; impartial and upright and, thus, keep a good management.*

In Summary

Be agile.
Get many opinions in forming your plans.
Keep your mind open.

Chapter 18

Take Advantage of Opportunity

The commander must create a helpful situation over and beyond the ordinary rules. By "situation," I mean he should act expediently in accordance with what is advantageous in the field and so meet any exigency.

—Sun Tzu

CAGEY NEGOTIATORS leave their options open so they can take advantage of the best opportunity. When Michael Dell is interviewed about his company's relationship with Intel (until their acquisition of gaming-computer maker Alienware, Dell used only Intel chips), he says, "We don't have an exclusive arrangement with Intel." That statement leaves the door open for Dell to have relationships with other microprocessor manufacturers and inherently works to keep Intel on its toes—and its price offerings to Dell competitive.

Every publicly held company needs to make its quarterly results. When negotiating for high-ticket items such as software, you can frequently improve the pricing you receive if you negotiate with a vendor who is having a subpar quarter. They will be much more amenable to giving you a price break to secure the sale. Accordingly, if you have flexibility in your timing or are talking to multiple vendors, probe and exploit the weaknesses that let you get the best price.

Sun Tzu was opportunistic in his operations. He advised, *If the enemy leaves a door open, you must rush in. Seize the place the enemy values without making an appointment for battle with him.* Sun Tzu took advantage of opportunities aggressively.

Taking advantage of opportunities means not overpaying in money, time, or the opportunity cost for your organization. You can avoid overpaying by constantly staying ahead of the curve on the areas you are looking at for opportunities. A current knowledge base helps detect and assess opportunities. Gut feelings and hunches play a part in business, but they must be backed up by research and knowledge of the marketplace. You can become a very good negotiator by doing a few things. Great buyers consistently follow a few guidelines:

They are always out testing the market. They meet with vendors they may not even want to do business with. This keeps their knowledge base constantly up to date. They know pricing—because they are always probing for a better price. They know about quality—because any good pricing conversation involves some quality discussion. They know trends, from multiple viewpoints, because they are always asking about what is new. This practice keeps great buyers constantly current on new developments in the marketplace.

They try to never get beholden to a vendor. Granted, that sometimes isn't possible. (No drugstore can avoid selling Crest toothpaste.) But very frequently you can exert leverage by saying no to certain vendors or giving better placement to their competitors. Great buyers keep as much of their business "in play" with as many vendors as possible.

They are loyal to people who have been loyal to them. Test the market and maintain your flexibility, but don't change vendors and relationships on a whim. If you find more competitive pricing and service, give your preferred vendor an opportunity to meet it. Also, look at the benefits of doing business with someone who already knows your business and expectations. Loyalty can translate into real dollars in terms of vendors who will go the extra mile for you on service.

Keep Customers as Your North Star

We all make money by serving customers. Taking advantage of opportunities in a customer-driven way best ensures a long-term payout from decisions made opportunistically. Being customer driven means that the customers' needs and desires are always taken into account in decision making.

Ancient mariners successfully traveled incredible distances using the North Star for navigation. If blown off course, they could get back on track using the North Star. The North Star didn't tell them where to go, but it did give them an unchanging reference point that kept them on course.

Customer-centered organizations use the customer as their North Star. Competitive jostling and changing markets may temporarily knock you off course. But you can always refocus by listening to your customers and what they want. At the end of the day, the customer—not your competitors, Wall Street, or industry pundits—determines who wins.

Customer-centered organizations share several characteristics:

Fact-based decision making. Facts can be market research or sales results. Fact-based decision making keeps emotion and opinions at a proper level. If used well, it will help keep your organization focused on the most important opportunities available.

Listening to the customer. There are different ways to listen to customers. Some companies have formal research programs or conduct regular focus groups. Some companies listen informally to customers—customer complaints can be an example. In retailing, some companies have a culture of listening to customers at the point of contact. This is easy and inexpensive to do as customers are available in the store at any time. Other companies dive deeply into customer complaints—and broadly circulate those (as well as customer testimonials) in their organizations.

Having a customer-comes-first culture. When decisions are made, are the customers' needs taken into account? This is a matter of organizational culture and style. In customer-driven cultures, talking about the customer is commonplace. And customers' needs are taken into account when major decisions are made. This is done through understanding the effect on the customer of a particular decision—how it impacts a customer negatively, how it impacts a customer positively. Many decisions do both.

Investing only where customers will pay for that investment. Great customer-centric organizations have a knack for putting their money where customers would want that money placed. They may invest in better-trained staff because the improvement in service is the most relevant—and immediate—improvement for the customer.

Some companies make profits solely on how they manage their money. Watching costs, maintaining margins, and keeping within budgets are their top priority. In customer-centered organizations, a different view of the world prevails—companies make money by serving their customers better than their competitors do. These companies understand that as long as they are favored by their customers, their financial future is solid. Making money in these cultures is the logical outcome of serving customers.

These companies still watch costs, margins, and budgets. Managing those aspects competently is part of managing any company. But these companies believe that financial metrics will fall into place over the long term when customers are well served. Customer-centered companies find a good balance between short-term financial success and long-term financial success.

Customer-centered organizations talk about customers at all levels, and take actions based on their perceptions of customer needs. At Procter & Gamble, employees are inculcated in a customer-driven culture. P&G believes that two different actions can grow their brands. First, better product—a consumer-relevant product advantage. Second, better communication of that advantage—a better connection with the customer. Those actions are viewed as the two ways long-term value is added to the organization. Other actions may add value, but not in the same long-term sustainable way.

Jeff Bezos of Amazon.com sums it up in the following way: "If you give customers what they want, the rest will take care of itself."

In Summary

Manage opportunities as a good negotiator: Don't overpay.

Keep customers at the forefront of your decision making.

Find metrics that disseminate customer-driven thinking.

Chapter 19

Use Surprise

Thus, one who is adept at keeping the enemy on the move maintains deceitful appearances, according to which the enemy will act. He lures with something that the enemy is sure to take. By so doing he keeps the enemy on the move and then waits for the right moment to make a sudden ambush with picked troops.

—Sun Tzu

IN BUSINESS AND in military operations, doing the unexpected can yield a significant advantage. The U.S. Army's pre-World War II advice in *Infantry in Battle* was, "Surprise is usually decisive; therefore, much may be sacrificed to achieve it. It should be striven for by all units, regardless of size, and in all engagements, regardless of importance." Because surprise is decisive, it is a significant asset in the business world.

FLEXIBILITY

Marketers talk about differentiating. Investment bankers go one step further and talk about being disruptive. That is how you destabilize a market and make a superior return, by disrupting the status quo. Surprise disrupts the status quo.

Surprise can take several forms. A new marketing strategy can surprise the competition—perhaps an unexpected lowering of prices, or a sudden increase in promotion. A particularly pointed advertising campaign can be a surprise—like a "Pepsi Challenge" comparison, where Pepsi is tasted side by side with Coke and wins because most people actually prefer the sweeter taste of Pepsi. An advertising execution like that can make your competitor rethink their core strategy. In the case of the Pepsi Challenge, it prompted Coke to introduce an ultimately unsuccessful reformulation of Coke that tasted more like Pepsi. Coca-Cola subsequently withdrew that product from the market.

Surprise forces your competitors to think in ways they hadn't previously. Hiring a key person away from a competitor can cause surprise and upset preexisting plans. That's why you will frequently see lawsuits when a company hires talent from a direct competitor. A new acquisition can upset a delicate competitive balance and cause an industry shake-up. And new products can recast market shares and force competitors to react, or retrench.

Innovate

Innovation is a form of surprise. It catches competitors off guard. And it gives smaller companies an opportunity to compete against more entrenched competitors. Few innovative ideas are completely and truly new. Most new ideas are a repurposing of an idea that started somewhere else. Wal-Mart borrowed the idea of a supercenter (a combination mass merchant and supermarket under one roof) from a Midwestern competitor. Ray Kroc bought McDonald's and simply had the vision to build more. Both of these innovations were young ideas that companies capitalized on aggressively.

How do you develop new innovation?

Talk to customers. Customer-driven organizations generally outperform organizations that don't listen to their customers. Many companies pay lip service to the idea of listening to customers. Listening to customers requires a willingness to take action on what they have to say—particularly when you don't like what the customer says or when taking action on the feedback costs you money.

Paco Underhill is a popular retailing consultant. In his study of shopping, he identifies a shopping difference between men and women. He says, "In one study we found that 65% of male shoppers who tried something on bought it, as opposed to 25% of female shoppers." That's an insight that can affect a store's layout, merchandising, and training programs.

Look at industries next to yours. Innovation frequently comes from outside of an industry. The founder of Amazon.com didn't come from the book business or retailing. Wal-Mart, the largest seller of food in the country, didn't come out of the food industry. Look at industries around you for ideas to apply back to your own.

Talk to industry watchers. Sometimes they are the press; sometimes they are consultants. These are people who talk to industry leaders and travel to see the latest innovations. These are people who will have a broad-based perspective on the industry.

Benchmark best practices. This is particularly helpful when looking at the "extremes." Who has lowest costs? Best service? Highest-quality products? What parts of these best performances can be brought back to your own company?

Listen to your own people. Within every organization is plenty of creativity. And plenty of ideas. Frequently from people who are close to your business—who have been thinking about how to solve problems on their own time.

In developing new products or services, one thing to look for is a "chord of familiarity"—a way a new idea is familiar to people who haven't seen that particular product before. A chord of familiarity gives people a way to understand a new product. For example, a battery-operated razor like Gillette's M3 Power or their Fusion is just a battery-operated version of a familiar, basic razor. Though it is new, it is still an understandable innovation, helping it to have a broad appeal.

Successful new ideas build unexpected value for a company and allow new innovations that build from the initial innovation. These can surprise competitors and dramatically increase the value of an enterprise.

In Summary

Surprise is disruptive to your competitors.

Capitalize aggressively on successful innovations.

Look inside for ideas.

Chapter 20

Lead Based on the Situation

Even though we show people the victory gained by using flexible tactics in conformity to the changing situations, they do not comprehend this. People all know the tactics by which we achieved victory, but they do not know how the tactics were applied in the situation to defeat the enemy. Hence, no one victory is gained in the same manner as another. The tactics change in an infinite variety of ways to suit changes in the circumstances.

—Sun Tzu

FLEXIBILITY

WHEN ASKED WHY his management style varies from individual to individual, natural foods executive Jeff Tripician says, "Because I want to optimize their performance. Since people are different, I get better performance from them if I manage different people differently." There's a benefit—"to optimize their performance." We don't manage different people and different situations differently because it's easier—frequently it isn't. We manage different people differently because if we optimize their performance, we optimize the performance of our team.

An example in the film industry is managing high-profile actors. Says the *Wall Street Journal* in writing about the success of leading directors, "Fine actors bring their own performances; they may require little more than the director's trust." Because of skill in that area, "That's why actors have been so eager to work with filmmakers like Robert Altman, Woody Allen, Sidney

Lumet or Martin Scorsese." These directors understand how to get great performances from great actors.

As a business manager, you have lots of tactics at your disposal to optimize performance:

Annual goals. Most companies tie pay and bonuses to sales, profit, and productivity goals. Use these goals to incentivize performance.

Weekly project reviews. You can use this tactic to keep employees on target with goals and expectations. As a manager, weekly project reviews can be helpful if you are new to the company, department, or trying to come up to speed on a particular project. Use this tool as the situation warrants it—for new team members or if a project has slipped off track.

Informal chats. Spend your time where you can make a difference—both informally and formally. While meetings are a necessary part of organizational life, you will often learn more taking time to have informal talks with your staff. The relaxed atmosphere will yield more candor than the feedback you get in a meeting. Meetings often have agendas. Informal chats are about gaining information.

Personnel reviews. Done well, these can be very helpful in setting future priorities. These should not be forums to cast blame or punish past actions; these are opportunities to direct future actions.

Build Processes Judiciously

Different companies succeed with different processes and management systems. In general, big companies have more processes than smaller companies do. That is often because the dollar risks are greater at big companies, thus driving more processes. You can build your processes to fit your company's unique size and needs by looking at the following factors:

Turnover. In high-turnover organizations, build well-defined processes to pass on company knowledge from one "generation" of employees to the next. This can be important in managing IT departments, or other departments where a booming job market can cause sudden turnover.

Experience of people. The more inexperienced people involved in your organization, the more processes you need. Well-defined processes define the steps in decision making so new people can step into a role and have guidance in making the right decisions. When new college graduates enter brand management at Procter & Gamble, they are given responsibility for managing the budget of several P&G brands. That's generally several million dollars worth of responsibility. But the processes and reporting are clear: every month, the new individual sits down with an accounting manager, and then subsequently a brand manager, to explain the current budget spending and any variances. No new, inexperienced employee would have been able to make too many mistakes without someone finding out fairly quickly. The well-set-up processes for these fresh-out-of-school hires ensures that.

The number of people involved. The more people that need to be involved in a decision, the clearer the process needs to be. Otherwise, decision making slows down or gets muddled. In particular, if decision making goes across multiple departments, simple processes can help make hand-off and project transitions run smoothly.

As you manage through fast-changing situations, some simple rules can keep you on track. These rules can vary with the situation.

An example is the nuclear incident at Three Mile Island. Harold Denton was then-president Carter's representative at Three Mile Island while the accident was occurring. He was in the difficult situation of determining risks and handling the press. And events were moving quickly. He needed to make a lot of decisions in a short period of time. Harold offers this set of rules for managing any crisis:

- Tell it like it is.
- Admit uncertainties exist.
- Don't make statements you will have to retract later.
- Act on the best estimate of a situation.
- Refrain from "value judgments."

President Carter said after the incident, "I went into the control room with Harold, and from then on I saw on television every night his calm, professional, reassuring voice letting the American people know that they need have no fear." Harold Denton's rules gave the foundation for better, quicker action. They helped handle the situation with both flexibility and decisiveness.

The more flexibility you keep in your management systems, the faster and more flexible your company will be. Build your processes to the specific needs of your company. Build guidelines and rules to help delegate decision making and ensure common criteria for those decisions.

In Summary

Manage to optimize performance of individuals.

Keep flexibility in your management systems.

Informality is a competitive advantage.

Chapter 21

Do Your Homework

Generally, he who occupies the field of battle first and awaits his enemy is at ease; he who arrives later and joins the battle in haste is weary.

—Sun Tzu

QUARTERBACK TOM BRADY is one of the NFL's best players. Brady does his homework before each game by watching films of the opposing team to learn their playing style and strategies. In an interview with *60 Minutes* in November 2005, Brady said, "A lot of time is spent . . . on the film and . . . trying to get as many pictures in your head before the game as you can. So when you do walk on the field, you can just verify what's going on. And it's not just to go back there and wing it. You try that, you are going to wake up Monday morning with headaches." In any business you do your homework, make your plan, and you go out and execute.

One of the world's most famous advocates of doing your homework is Warren Buffet. His philosophy is to buy into only those businesses that you can understand. Buffet has never been a big investor in high-tech companies; he claims not to understand them as well as he does other businesses. But he has done well investing in businesses that he does fully understand. He says, "When you are convinced of a strong business's prospects, be aggressive and add to your position rather than buying the

15th or 20th stock on your list of possible investments." That's the confidence that comes from doing your homework. Like Warren Buffet, when you strongly believe you have a winner, go with it.

Homework in the business world takes a lot of forms:

Practice best practices. Tom Brady isn't the only NFL quarterback who watches game films as part of his preparation. Many of the NFL's best quarterbacks are particularly noted for being game-film junkies. Know the best practice in your industry, and make that part of your routine.

Get into the field. Every job has a version of this. If you work in a corporate office, go on sales calls with your salespeople or account managers. A common part of any job, particularly if you are in the corporate office or executive suite, should be to get out to talk to your customers. Go to where your products are sold and talk with the people that sell them. This is where you learn the effects of the decisions that you and your peers make. Great organizations listen and respond to what they learn about in the field.

Do as much of this as you can with your competitors' products as well. In retailing, a common part of management positions is to visit your competitors' stores and see what they are doing. Every industry has a version of this—get out and get as close to your customers, *and* your competitors' customers, as you can.

Study your business. Digging in and understanding your business will always take some work. Make time to understand how your business is doing. Use your internal reports, use industry data, and use people who know about your business.

Learn related aspects of your business. As you progress through your career, you will frequently work in areas that are unfamiliar to you. However, a new product, acquisition, or strategic direction can create an opportunity to work in a part of the business unfamiliar to you. Use these opportunities to learn—it's good for the business and good for your career. The more you learn about the business around you, the more you can make your part of the business (1) respond to the needs of those around you and (2) fit into the corporate needs.

Learn in different ways. We all have preferred ways of learning. This is good up to a point. But to broaden your base of knowledge, you need to learn in different ways. Many people spend most of their learning time with reports. However, you can also learn from:

- Customers
- Vendors
- Peers
- Your supervisor
- Subordinates
- Industry experts
- Conferences and industry gatherings
- The competition

The best learning comes from your devoted customers. Because of their frequency of interaction, they likely know what needs to be fixed or improved about your company. And since they are your best customers, they already like your product, service, or company and want to see it succeed.

Make Preparations

In Sun Tzu's writings, there is no theme on execution that comes out more clearly than preparation. Sun Tzu writes, *Materials for setting fire must always be at hand*. And he says, *Know the weather and know the ground, and your victory will then be complete.*

Sun Tzu's preparation was thorough. He writes, *If in the neighborhood of your camp there are dangerous defiles or ponds and low-lying ground overgrown with aquatic grass and reeds or forested mountains with dense tangled undergrowth, they must be thoroughly searched, for these are possible places where ambushes are laid and spies are hidden.* In Sun Tzu's world, these are the "basics." Take pride in getting the basics done well.

In the business world, preparation takes many forms. Training programs are preparation. An annual marketing plan is preparation. Doing your homework for a big sales call is preparation (as is arriving in town the night before a morning meeting—instead of taking the first flight out that same day). A business continuity plan (a disaster plan) is a form of preparation.

Preparation is about looking at the big picture, but it is also about managing the details. Writer William Feather once said, "Beware of the man who won't be bothered with details." In business you need to know the metrics of your business—what percentage of market share you (and your competitor) has, how much your revenue increased each year over the last five years. Every business has a different set of details that need to be managed well.

Preparation goes hand in hand with flexibility. Good preparation gives you the skills and materials and the opportunity to be flexible on the battlefield. As a situation changes, you can take advantage of it. This holds true in the business world. Being prepared and flexible is a good formula for winning.

In Summary

Make learning part of your daily work.
Study your business.
Learn related areas.

Chapter 22

Get the Message Through

Do not allow your communication to be blocked.

—Sun Tzu

THROUGHOUT HISTORY, people have gone to extraordinary lengths to make sure their message—and its meaning—gets understood. In ancient Greece and Persia, a technique for sending communications was to shave a messenger's head, write a message on his scalp, and wait for the hair to grow back before sending the messenger off. Only upon shaving the hair off of the messenger's head would the message then be visible. That way, if the messenger was captured, the message would be hidden under a head of hair and not discovered. With the advent of electronic communications, the importance of cryptography—ensuring that a message couldn't be read if the message itself fell into the wrong hands—increased. Stories about Allied code breakers who could break the German Enigma code and the Japanese military code have become a great part of our World War II understanding in recent decades.

In business, many things can affect our ability to get a message through. As a "sender" we may not be an effective communicator. Because we are human and thus experience messages through our perceptions and experiences, communication is more an art than a science.

If you want to create change, you need to communicate. There are different techniques to use to be an effective communicator:

Influence. Teddy Roosevelt described being president of the United States as a "bully pulpit." It was a place he could influence "the people." One of the key strengths of the office for Roosevelt was how it helped him influence and communicate with everyday people.

Influence is more important than commanding people who report directly to you. Being able to broadly influence those people around you is an outstanding way to *support* people who work for you. Consider the human resources department in many organizations. HR is an area that has contact with a lot of different departments and with the public via recruiting new employees. But in many organizations, its clout is generally low—it leads without a lot of direct reports. It doesn't set budgets or create new products or accounts. However, it does have a role as an influencer or mediator in many internal company issues. It dispenses advice to employees at all levels of the organization. In many organizations, a strong HR department gets its strength from its ability to form bonds with people, to develop trust, and to guide outcomes. It may not hold strong positional power, but it holds power in its ability to influence.

Find networks. Networks of people disseminate information among the group. Says one Buffalo, New York, native about the arts community there, "People in the arts, for example, all have connections to one another. This arts community may be small compared to some others, but it is one of the least compartmentalized I've ever seen." Networks like this can spread *relevant* information quickly.

Look for allies. Who can help you spread your message? Most likely, someone who believes that message as well. Look for fellow believers. In any organization, there are people who share your point of view. Maybe you are new to a company and there are other new people who have the same perspective. Perhaps there are people faced with similar business challenges. People who share your opinion without any convincing can be valuable allies.

If a project or initiative is important to you, have a communication plan. You probably already have a project plan or a timetable. That's a good start. But make sure you know how

1. To get informal feedback on your project (that's where you find out how things are *really* going),
2. To continue building organization support for your initiative, and
3. To report results.

When it comes to reporting results, don't naively assume you can report good information in the same manner you would bad information. Bad information needs to be managed. That doesn't mean you should hide it—that's a practice that can endanger your project and your reputation. However, reporting good information, good results, requires less information. Bad information needs to be reported in person; good information can be more formally disseminated.

It's about the Style, Too

How often have you heard this: "I sent out an e-mail and I didn't hear anything back"? E-mail is a very efficient form of communication. The sender can immediately communicate with a broad group of people. Respondents can respond at a time that works for them. However, in part because of its efficiency, it is frequently abused through overuse. Be personally effective in how you communicate. It is important because it affects people's perception of you.

Use e-mail appropriately. If something is really important to you, visit or call the appropriate person. In a world of excessive e-mail communications, a visit or call emphasizes the matter in a positive way. And don't forget old-fashioned communications. Because they are more infrequently practiced, they may make even more impact now than they did years ago. A handwritten note stands out for the recipient. A senior leader going out of his or her way with a handshake and a thank-you is even more memorable.

Managing in Difficult Times Is Even More Challenging

Managing in a company that is experiencing a difficult time requires a greater emphasis on communications. That's because fewer of your company's communications will be in your control. If your company is big enough, quarterly results or major announcements (a downsizing?) will be posted in the paper.

Managing in good times is straightforward. In difficult times, take control of the flow of information:

- Let your employees hear news from you or your company before they hear it elsewhere.
- Let people know what is next.
- Communicate bad news as aggressively as you communicate good news. Stop the rumor mill before it starts.

In Summary

Communication takes time: Plan accordingly.
Go to whatever lengths you need to in order to communicate effectively.
Vary your communication style with situation and the group.

Have Towering Strengths

Introduction

Be Great at Something

COMPANIES WHO execute well work at it. If that execution is consistently focused on building competitive advantage, a company can develop towering strengths—strengths that propel the business to a high level of competitive advantage. These strengths come from building one strength on top of another. These strengths can eventually add up to a compelling long-term advantage. These strengths accrue to companies that have invested, and continued to invest, in getting good at a particular aspect or benefit of their business.

Target used to be just another mass merchant. But its lineage descended from a department store—Target's parent company owned Marshall Fields, among other department store brands. Over time, Target brought style and fashion to a mainstream market better than anyone else did. This wasn't an overnight transformation. Target started with designer Michael Graves in 1998 and has added additional designers since then, spreading "design" across its stores and throughout its various departments and products. Now Target designers encompass everyone from Sean Conway for gardening to Sonia Kashuk for makeup.

Intel didn't always make microprocessors. It started by making random access memory circuits. But in the 1980s, the memory chip market was becoming crowded and commoditized. Intel's shift into microprocessors allowed it to take a leadership position in a (at the time) younger market. Now Intel is the microprocessor gold standard—even causing Apple to abandon its G5 microprocessor and move over to the Intel camp.

Japanese car companies' first cars in the U.S. market were cheap—and not of very good quality. Forbes.com nominated the 1978 Honda Accord hatchback as one of the worst cars of all time. But the low prices were enough to achieve sales success and a toehold in the marketplace. Using modern quality-control approaches, the Japanese managed to rapidly improve quality while U.S. companies slipped behind. Now the Japanese have Detroit car companies on the run with their combination of quality and value. Toyota, in particular, has established leadership in the midrange price segment with various vehicles and a strong presence in the upscale segment with their Lexus brand.

All of these companies are now demonstrable leaders in their fields. They have in one way or another superseded the competition. Maybe in costs, maybe in how they do business, maybe in the image they hold in the customer's mind. Each of their towering strengths allows them to build their business at a faster rate than their competitors can.

Thinking about building towering strengths is a different exercise than most business conversations. A lot of our daily efforts in business are focused on improving things that went wrong. Maybe selling costs were too high for a month, maybe a sales projection was missed, perhaps receivables were too high.

Creating towering strengths requires:

Investing in areas you want to get better at. If you aren't willing to invest a little, you (or your company) doesn't want to excel in that area. In bigger organizations, align yourself with trends or business directions the company is already interested in. Alternatively, find a mentor and low-cost ways to demonstrate a payout for your ideas. Keep the customer as your lightning rod, and your chances of succeeding increase.

Picking areas you can beat your competitors at. Attacking a competitor head-on doesn't work as well as attacking at a vulnerable point. Southwest, JetBlue, and other discount airlines won by building a cost structure that was different than the larger airlines'. They started by winning over the leisure traveler, who was less influenced by the larger airlines' more convenient airports or frequent-flyer programs, which had won over the business traveler.

Knowing what your company can be good at. Wal-Mart's business is based on hiring lower-skilled, lower-wage employees at their stores. This allows Wal-Mart to do a lot of things for low cost. But being able to perform complicated or skilled tasks at the individual store level is not something that Wal-Mart is currently strong at.

Know what your customers want. For a company selling to Wal-Mart, thinking this way can make you a success too—align your thinking and your strengths with the strengths of companies you are looking to sell to. If you sell flowers, package them to move through Wal-Mart's logistics and to the individual stores with minimal handling.

Defining a new standard across your company. Few of us accomplish anything important by ourselves. Spreading the understanding of what a company's competitive advantage will be is as important as defining that advantage. Market research can be a tool to define the competitive, or customer-defined, opportunity that you want to fill.

Towering strengths give you an outsized advantage over your competitors. This can cause your customers to forgive an occasional mistake because what you offer can be difficult to find somewhere else. "Forgiveness" from loyal customers—forgiveness when the company makes a mistake—is one quality of a strong brand.

Chapter 23

Create Your Own Strengths

During the process from assembling the troops and mobilizing the people to deploying the army ready for battle, nothing is more difficult than the art of maneuvering for seizing favorable positions beforehand.

—Sun Tzu

SUN TZU'S "seizing favorable positions beforehand" is the military version of a business building its own unique strength. For Sun Tzu, a favorable position gave control over the battlefield. He dictated the terms of engagement—or the enemy attacked and Sun Tzu was in an advantageous position. Building unique strengths provides similar advantages to the businessman. Unique strengths can come in a variety of forms:

- A position in a customer's mind
- A business model
- A series of capabilities no one else has developed
- A low-cost structure that is difficult to emulate

If you don't build a towering strength, you are vulnerable. Any traditional retailer competing with Wal-Mart knows this—"The move to value and premium will accelerate and the middle will become even lonelier," says John Lovering, chairman of British retailer Debenhams. The world of retailing is crowded. Creating a towering strength is challenging. What's more challenging is that you have to make money as well.

Low-priced retailer Costco is different from other retailers—and more successful. How does it manage success with low prices and low margins? Costco's is a unique formula:

Make sure each sale is large. You won't find $1 or $2 items at Costco. For example, there are no individual candy bars in the checkout lane. Soup is available only in a six-pack, and beer is available only by the case. By keeping virtually all sales above $5 a package, Costco can make its low-price and low-margin model work. Costco makes sure it makes enough money on each sale. Costco customers just have to buy more at a time to get the savings.

Carry a limited assortment. Each Costco carries no more than 4,000 items. The average supermarket, as one benchmark, carries over 30,000. At Costco you will not find three or four brands of canned green beans—you might not even find canned green beans. At mass merchants, you will find several microwaves to choose from; at Costco, you will generally find two or three—a big one, a small one, and sometimes a medium-sized one. But Costco members have come to trust Costco's selection—Costco is a trusted buying agent.

Only carry the highest-volume items. With only 4,000 items in a store, you have to be selective. And Costco is known for carrying only items that move quickly. It is also known among its vendors for quickly discontinuing items that move slowly.

These guidelines are carried out consistently by Costco. Over time, its model is understood and makes sense to its customers. No Costco customer expects to find a large selection. But customers also know what to depend on Costco for—and what they can depend on Costco for better than almost any other retailer is value on quality items. In a playing field crowded by mass merchants, supermarkets, and home improvement stores, Costco has built a towering strength on value. It may lose on variety. It may lose on convenience (the lines are almost always long at a Costco). But its customers know that what Costco does, it does well.

Without a unique strength, you are at risk to be in a vulnerable middle-market position. In the "hourglass economy," customers are saving money on basics where they can—at places like Costco—and using the savings to trade up in other areas. Middle-market brands like Kraft and Ford lose because they are neither the staples we buy at good prices nor the brands we trade up to. Kraft loses to Costco's or Wal-Mart's private label at the low end and imported European or artisanal cheeses at the high end. Ford loses to less expensive or more expensive imports.

Starbucks has redefined the coffee market in the last few years. Coffee used to be dominated by Folgers, Hills Brothers, and Maxwell House. It used to all come in cans. Remember Mrs. Olson? What could be more contrary to the current trendy image of coffee? (Mrs. Olson was an older female character in Folgers commercials for twenty-one years—back when mass-market brands ruled.) Today, coffee is more expensive, more hip, and can be fairly complicated, depending on what you are ordering at your local espresso bar.

Folgers used to be "good to the last drop." Folgers had a taste-oriented strength that years ago could have given the brand leverage to transition to an up-market brand before Starbucks established that positioning. But that opportunity has passed as Starbucks clearly owns the upscale coffee market—and "legacy" brands like Folgers either fight it out in the middle or look for opportunities as cut-rate basics.

Despite missing the chance to go up-market in coffee, in many categories, Procter & Gamble has achieved years of growth through a focus on the customer. In more traditionally stable categories than food, where P&G's research and development strength can be leveraged, P&G has created new markets and new standards by harnessing its research and development to a focus on unmet customer needs. This core strength has let P&G innovate trade-up products like the Swiffer dust mop in recent years.

Look at the staying power of P&G's brands in disposable diapers. Pampers was introduced forty years ago! Towering strengths, once created, can provide sales and revenues for a long time.

In Summary

Execute consistently over time.
What you don't do also defines your strength.
Defend your position.

Chapter 24

Invest for Long-Term Advantage

Ground that both we and the enemy can traverse with equal ease is called accessible. On such ground, he who first takes high sunny positions and keeps his supply routes unimpeded can fight advantageously.

—Sun Tzu

BUILDING LONG-TERM advantage requires some investment. True long-term advantages are sustainable—there is an aspect of them that makes them difficult for a competitor to exactly replicate, and difficult for a competitor to compete with. But once you have it, you can stay miles ahead of the pack. Or, as Sun Tzu says, you "can fight advantageously."

There are a variety of ways to build long-term advantage.

In retailing, some companies build long-term advantage by their real estate strategies. Some companies buy the best real estate and use that as their advertising—a timeless sort of advertising. These companies follow a real estate strategy called "Main and Main." This refers to rural small towns that invariably have a Main Street that is their busiest street. A location at the corner of Main and Main would be a very busy location! So these companies look for real estate at very busy intersections—the hypothetical "Main and Main"—real estate that is very expensive because of the high volume of traffic that goes by it. This strategy trades the higher cost of premium real estate for a long-term competitive advantage.

In consumer packaged goods, some companies build a long-term advantage through their patents. However, this also can be very expensive. For every product that becomes a successful patent, many products are developed that never get that far. An expensive winnowing out takes place.

In a variety of industries, like credit cards, companies build long-term advantage through their customer databases. These databases take a large amount of technology and experience to operate well. Yet, they are an important tool for retaining customers and building sales significantly.

Consistently, companies that build a long-term advantage invest in one aspect of their business at the expense of other areas. Singapore Airlines is known for its gracious service and dependable on-time performance. Perennially, it is one of the favorite business airlines in the world. It got there through its strong focus on serving the customer. Says one analyst, "On everything facing the customer, they do not scrimp. On everything else, they keep costs low." An example of scrimping where their customer is not affected is Singapore Airlines' modest offices—they are on the top floor of one of their hangars.

Your Long-Term Advantage Is Not Always Obvious

For years Sears has invested in brands like Craftsman and Kenmore. As Sears sunk over the years, these brands remained solid. Even in Sears's worst year, advertising for Craftsman tools still appeared on TV.

When Sears and Kmart merged, there was much speculation about rebranding Kmart stores as Sears stores. But the tarnished Sears brand wasn't much stronger than the tarnished Kmart brand,

and those changes have had uneven results. However, Craftsman and Kenmore appliances sell well at Kmart. The strength of the Sears brand isn't the retail stores or the in-store experience tied to that brand. Actually, the strength isn't the Sears brand itself at all. The towering brand strength at Sears is the in-house, "own brands" that Sears built over the years.

Build a Unique Capability

Achieving a long-term advantage takes some work and planning. To build a long-term advantage, you need to be better at something than your competitors are. And ideally, make your strength, your advantage, something that is difficult to copy. As Gerry Hodes, formerly of Marks and Spencer, says, "Everybody wants to pick the low hanging fruit." Get good at something that lets you pick the higher-hanging fruit—fruit that is difficult for your competitors to reach.

Build a long-term advantage that:

Capitalizes on an aspect of your company that is unique. By doing this, you make it difficult for your competitor to copy you. If you are a small business, tying to your local geography can make you different from national competitors who can't compete as nimbly locally. Garlands supermarkets in Los Angeles does this with a uniquely Californian décor scheme—reinforcing its local roots and ownership.

Builds something that takes time to get good at. Anyone who has implemented a high-tech customer management system will report that the first step is just the first step. There are many steps to follow. Learning takes time. First, you learn your leverage points—what you can tell or communicate to your customers that they will react to. The process of "learning" takes years.

When it is done, companies from Nestle to Tesco report, what they *know* about their customers, and how they can take *action* on that knowledge, is a unique strength.

Leverages a strength from another part of your business. Acquisitions are one way to do this. When Yahoo bought online photo-storing site Flickr, it was for the knowledge about "social media," practiced by Flickr, that Yahoo could transfer to all of its organization. When $30 billion in sales, Albertsons bought Shaw's markets, which was $3 billion in sales. Albertsons took Shaw's loyalty-card marketing practices and spread those back through the entire Albertsons organization. Those marketing practices were far more advanced and effective compared to what Albertsons was practicing at the time.

Cost can be a unique capability. But if it is easy for competitors to copy, it won't survive. Frequently, companies with a strong cost advantage develop it across a couple of dimensions. That adds to the competitive insulation. Even so, cost-based capabilities are among the most transparent and generally require scale to deter copycats.

Southwest famously started the low-cost airline industry. They offered a significantly simplified operation—a standardized fleet of planes, no first class, and no reserved seats. And they added nonunion, lower-wage people (who seemed to actually enjoy their jobs). But all of that could be equally said now about JetBlue as well, and a variety of smaller low-cost airlines like Frontier. Others followed the same formula.

Similarly, dollar stores (Dollar General, Dollar Tree, 99 Cent Stores, etc.) all take advantage of

1. Cheap production in China (a highly disproportionate share of sales comes from Chinese manufacturers)
2. Lower-rent strip mall or rural locations, and
3. Minimal in-store labor.

They are a strong competitor to many other retail formats, but they also tend to be highly interchangeable and cannibalistic with each other.

Take Advantage of Changing Markets

For many, success has come from taking advantage of a strength in a way they had not originally planned. Cisco grew to be the world's largest maker of Internet networking equipment as the Internet itself grew. This strength was built on the business-to-business market. But as the Internet grew, the consumer market for Internet networking also grew. First, in wired houses for people who had the means to afford that luxury. But like many stories in electronics and technology, the technology for Internet networking a house came down in price. And as that became wireless as well, the home market grew significantly.

Cisco, the business-to-business provider of this service, decided the home market was its next opportunity. Said Cisco, "Consumer electronics companies have been able to compete on a stand-alone device but the dynamics of the market are changing. The Internet and new networking requirements are enough of a disruptor for us to enter a new market." Said one analyst, "The home network is the last piece of territory up for grabs in the networking space and it makes sense for Cisco to try to dominate that." Cisco's success in the business-to-business space has prepared it for possible success in the home market as well.

PetSmart is the country's largest pet-food supply store. Its history has several twists and turns. In the 1990s, it was a fast-growing company with good margins. It had a popular line of products. In particular, the Hills Science Diet and Iams pet food brands were strong draws. These brands were only available through the "specialty pet-food channels." That meant veterinarians, mall-based pet stores, smaller pet-food stores, and the chains of pet-food stores—of which PetSmart was the largest.

That changed. First, Hills Science Diet began to experiment with selling its product through supermarkets. Then Iams was bought by Procter & Gamble. Procter & Gamble wasn't aligned with the specialty pet-food channels and quickly expanded the brand to mass merchants and supermarkets. PetSmart's biggest advantage had vanished. The stock price fell significantly. Management changed.

Many people treat their pet like a member of the family. So PetSmart and the category continued on. After all, the specialty pet-food retailers let you bring your animal into the store with you to do your shopping. And the retailers in this channel eventually built private label and other offerings that let them rebuild some margin.

But the greatest innovation was in services around the core strength of caring for your pet. PetSmart started by building veterinarian centers in their stores. Similar to how LensCrafters and others, years ago, began to compete with the local ophthalmologist by putting eye doctors in their own stores. Vertical integration, some might say.

Now, PetSmart has over 400 veterinarian centers in their stores through their partnership with Banfield. Pet grooming is another area that PetSmart has grown into. Virtually every PetSmart has a pet grooming center. And most recently, PetSmart has gotten into animal boarding. These pet hotels are also located right in your local PetSmart, and are cleverly marketed to have "all of the comforts of home." And, of course, you can buy extra services for your pet while you are away. An extra walk. Maybe an extra treat. All to soothe your mind that your pet is happy while you are soaking up the sun. In just a few years, these PetsHotels have expanded to PetSmarts in fifteen states.

A classic "consolidation" strategy was used—taking industries (veterinarians, pet grooming, and pet boarding) where there was very low market consolidation and using a strong brand to establish leadership.

This worked because, as the pet-food market changed, PetSmart had a clear strength in a brand that was trusted by pet-food owners. And it leverages that strength as it continues to build its business from just selling products to selling services as well. These and other services have grown to just under 10 percent of PetSmart sales, from just 5 percent in 2001. And these services contribute to a margin growth of almost 1 point per year. Long-term advantage in money-making segments should be particularly valued. If you are a small or medium-sized company, identify emerging niches and gain an advantage in them early. Ride these niches as they grow. Long-term advantage is most efficiently built with time, and long-term trends, as your ally.

In Summary

Build one rock-solid strength.
Anyone can pick the low-hanging fruit.
Leverage your strengths in new ways.

Chapter 25

Experience Matters

Those who do not know the conditions of mountains and forests, hazardous defiles, and marshes and swamps cannot conduct the march of an army. Those who do not use local guides are unable to obtain the advantages of the ground.

—Sun Tzu

FOR MANY great leaders, their greatest victories came at the end of their career after they had gained a lot of experience. Trafalgar was Nelson's last battle and his greatest victory. Most World War II generals had military experience in World War I; many U.S. Civil War generals had experience in the Mexican-American War. Before getting to a big stage, most successful leaders have honed their skills on a smaller stage.

Experience matters. In the United Kingdom, Tesco has come to dominate mass merchant and food retailing. It is bigger than Wal-Mart's UK entry, Asda. And it maintains a strong leadership over the other leading food retailer, Sainsbury. And Tesco's historical rate of growth has been so strong, industry watchers wonder how much more business it can take from its competitors. The UK press occasionally wonders if Tesco is on the verge of becoming a monopoly in British retailing. Tesco has done this under one leader and with a stable management team. On the other hand, its competition has been through a parade of managers with a variety of backgrounds—four at Sainsbury from 1998 to 2005, and four at Asda from 1999 to 2005.

Look for the Right Kind of Experience

A former boss of mine used to ask the following question regarding experienced job candidates: "Does he have 20 years of the same experience, or 20 different years of experience?" Essentially, doing the same thing over and over, in the same way, doesn't teach you anything new.

Great experience for developing leaders is to be in charge of something or to own a process. Some jobs inherently have this:

- Store manager jobs inherently give a breadth of experience. At some big box chains, a store manager runs a $75 million per year business. And managers handle the same issues as any CEO—sales, costs, and HR. These jobs can give people good experience, sometimes early in their career.
- Chef jobs can give similar experiences as they are responsible for a staff of people, the quality of the product, and costs. Chefs must also have a bit of an accountant and market researcher in them as they develop future menus—menus they use to manage costs and deliver the right profits.
- At a higher level, division manager jobs give a broad and full perspective. These jobs are responsible for the short- and long-term success of the business.
- Owning and operating your own business, even at a small level, can build ownership. To pay his way through college, financial consultant Steve Doner operated a mobile DJ business. He bought equipment, did sales and marketing, and worked the weddings and parties that he booked. With each decision, he tried to figure out how to make a little more money at the business. It proved to be great experience

for the decisions required in building his future financial consulting business.

These are all jobs in which daily execution is important, but in which short-term execution is also done in the context of longer-term responsibilities. Successful leaders know how to manage a variety of priorities—they know how to keep multiple balls in the air. They also know how to manage risk. Jobs that let people fail on their own in small ways are great for teaching people how to balance priorities and manage risk. People who have had jobs like these are good candidates to promote and to recruit. Their prior experiences give them a chance to be very good at execution.

Find Leaders with Passion—Let Them Build New Strengths

Lee Iacocca said simply, "I loved cars. I couldn't wait to go to work in the morning." Sometimes with hard work and experience, deep passions can develop. Many of the *Fortune* 100 Best Companies to Work For in America are privately held companies. In many of them (Timberland, Wegmans, S.C. Johnson, to name a few), the first-generation family member isn't running the company anymore. The baton has passed to seasoned people who, literally, grew up in their business and developed a passion for keeping it successful and growing.

Says Oprah Winfrey, "What I know is, is that if you do work that you love, and the work fulfills you, the rest will come."

Frank Perdue worked hard to make a great chicken. He developed proprietary breeds. But more than anything, he seemed to truly believe he made a great chicken. Frank Perdue talking about his chicken on TV is a timeless piece of advertising. What made those ads work was Frank. Says Joe Nocera in the December 25, 2005, issue of *New York Times Magazine*, "What comes through most of all in those ads is that whether or not his chickens were better than his competitors, Frank Perdue believed they were."

When food merchant, and former chef, David McInerney is involved in product development, he has a lot of people taste the new food products he is working on. When someone likes the item, he frequently asks, "Did you like it or love it?" David is looking for the products you love. It is at this early point in execution—when the product is still in development—that a lot of work goes into making sure that the product being developed is one customers will love.

This makes a lot of sense. Customers will go out of their way to get a product they love. And if they buy that product from you, maybe they will buy something else at the same time. Think of your favorite restaurant. There is probably a particular dish there that you love. That's what brings you there. And you likely don't go there alone. Maybe you bring your spouse along. Perhaps you meet friends there. That one item you love brings the restaurant a lot of business.

Products you love also lead to strong word of mouth. Any advertising person will tell you word of mouth is the best advertising that there is. None of us tell our friends about products that are just okay. We tell people about items we have developed a passion about. When we feel that passion, we tell people.

This is the way great brands grow. You could almost call it "organic"—though frequently strong word-of-mouth brands have teams of people working to keep that word of mouth going.

Harley-Davidson is a great brand among its customers. Harley-Davidson customers talk about their motorcycles. Emeril Lagasse is a great brand—his fans talk about his show and visit his restaurants and buy Emeril-branded products at their local store. Starbucks isn't a big traditional advertiser. But Starbucks fans will seek out Starbucks coffee whenever they are traveling.

Remember: Passionate customers aren't shy—they will talk about your product and service. Why? Because advertising only goes so far—their experience with your product or service is what matters most.

In Summary

Stability of talented leaders is a huge advantage.
Hire people who have "owned" something.
Build experience in something you have a passion for.

Chapter 26

Make Your Work Matter

What can subdue the hostile neighboring rulers is to hit what hurts them most.

—Sun Tzu

A CRITICAL TOOL in making your work matter is defining what advantage you want to build versus that of your competitors. Market research is a key tool behind this—it doesn't necessarily have to be expensive. Market research can help you do this by identifying gaps where you can perform better than the competition.

Most market research looks at shortfalls and short-term changes. And businesspeople go to work fixing what isn't working well. Momentum starts moving in that direction with the thinking before the research. Retailers look at basic service scores. Consumer packaged-goods companies look at price image and product performance. And most of this stems from a traditional view of the business.

Building a towering strength starts with a view of what strengths could be built, what competitive advantages could be grown. That leads to market research questions that are different from the norm—and a willingness to ask some questions in advance of a company's emerging competence. Ultimately, this pays off in building strengths that are different from the competition's—strengths that are uniquely yours.

Good market research breaks down results into *actionable* segments. That means that the market research ties to the organization of the business. If your company is organized by sales territory, the market research should be organized by sales territory. This allows accountability for the results of the research, accountability that is easy to "enforce"—the results will speak for themselves. Businesspeople who want to be successful, and want their business to be successful, will take action to make their business better. The only way you make more money by spending money on marketing research is if the results are organized in an actionable way—and the more people who can take action based on those results, the greater and faster an organization will make the desired changes.

To use market research to build a towering strength:

Write down who your target customer is. Who does your company want to buy your products? The more clearly your target is defined, the more relevant will be the strengths you build. Clear definitions of target customers inherently exclude some people. If you are not excluding some current customers, rethink the level of targeting you are building.

Identify the gaps you might want to build. What can your company be better at than the competition? What could you own? Where do your strengths seem to match some products or services customers want? What are they already buying from you more then they do from some competition—how do you turn that into a unique strength?

Ask customers what they think. And use this process to narrow down the gaps you want to develop. Focus on two key issues: Are your potential gaps, potential future strengths, important to customers (would anyone care if you were the best in these areas?), and Where do you stand versus your competitors? Ask enough questions, in enough different ways, that you can see the issue clearly. Bracket it with different questions to ensure you gain a clear understanding.

Track your performance against filling that gap, and communicate those results across the organization. Results will come slowly at first. But by tracking your results and improvements, an organization can see—and get excited about—the results.

Celebrate success. Focus, finish, and celebrate. Every team wants to win—and wants to beat the competition. Marketing research gives you a way to celebrate some wins. And to motivate a team to go out and win again.

This process works for executing change, and creating new strengths, for two reasons. First, it inherently breaks down work into actionable pieces. As you go through this, you will assign responsibility for building new strengths—ownership. People will take pride in this work—it will feel important to them.

The second reason this process works is because it lets you see the progress against building strengths. Progress will be uneven. If you work to create multiple strengths, you will be able to see the first successes. Use these successes to make people feel good about their work and even more committed to building new strengths.

No industry is easy. So as you work to build strengths, look at future trends you can own. Look at the margins of the industry. And look at industries near yours. Generally, you will not find any *one* strength that you can build that will trump the competition's. Frequently, you will find a series of small strengths that *deeply* matter to a few people. Building several of these strengths ends up mattering to many people—and a strong business is built.

Don't Hit Your Competitors Head-On

Hitting competitors where it will hurt the most is not hitting them head-on at their strengths. That won't hurt them.

- You don't hurt American Express by attacking them on service—they own that ("membership has its privileges"). But you might hurt them by attacking them on breadth of acceptance, annual cost, or with a different rewards program.
- You don't hurt Domino's pizza in the pizza-delivery business by telling people you arrive on time. People know and trust Domino's delivery service. But you might hurt them by coming up with a better pizza and telling people about that product. That is the attack Papa John's has tried in this business.
- You don't hurt Maytag by telling people your appliances work well—they built reliability. But you might be able to undercut them on price and value.

Sun Tzu sums it up, *Now, the art of employing troops is that when the enemy occupies high ground, do not confront him uphill, and when his back is resting on hills, do not make a frontal attack.*

In Summary

Ask many questions on your journey.
Commission many in your organization to help.
Use intelligence to build constant feedback.

Chapter 27

Redouble Your Efforts

In difficult ground, press on.

—Sun Tzu

LITTLE THAT IS WORTH winning comes easy. Steve Ballmer of Microsoft says, in the February 2001 issue of *Fast Company* magazine, "I like to tell people that all of our products and business will go through three phases. There's vision, patience, and execution." Patience is the time when persevering is most critical.

Don't Expect Wins to Be Easy

Mike Eruzione, the captain of the "Miracle on Ice" hockey team of the Winter Olympics, is now a motivational speaker. He's got a great story to motivate with. That team's victory in the Olympics has been called the greatest sports upset of the century. But it didn't get off to a good start. The opening game for the U.S. team of that Olympics was against Norway—a team the United States was supposed to beat. That game ended in a disheartening tie. Coach Brooks made the team do an hour and a half of intense skating. From the backboard to center ice and back again for an hour and a half. At the end, Coach told them if they didn't win the next day, they would have to do that again. The team won that game 8 to nothing.

Mike sums up his Olympic experiences with, "People who are successful make sacrifices." That 1980 "Miracle on Ice" team trained like no other U.S. Olympic Hockey team before them. That team practiced together for six months before the Olympic games.

Figure out what you will give up to be successful, but don't lose balance with family. Less golf, giving up a couple of nights of TV, getting up earlier every morning are all appropriate sacrifices to make. These are the kinds of conscious trade-offs that, if you are willing to make them, can free up time to invest in future success.

Beware Lurching

Many publicly held companies dash into a new strategy with great enthusiasm. Generally, it is this year's management trying to make their mark. It is almost predictable:

Auto companies will swear off their dependence on rebates every few years. And after a short time, they will return to rebates (they trained their customers to shop that way, after all).

Competitors of Wal-Mart will say, "There is more to shopping than price," and then fail to offer a shopping experience or product selection that offers a better–than–Wal-Mart alternative (after all, developing a point of difference to Wal-Mart requires vision, then patience—and many publicly held companies have a difficult time sustaining either).

Manufacturing companies will declare they are "customer driven," and then will introduce "me too" products that don't stand the test of time (15,000 new food products are introduced each year, as just one benchmark).

These are generally not strategies that shift execution. So they don't stand the test of time. Generally, the next management team signals some different, equally obvious, shift. And these kinds of shifts can happen at all levels of an organization—at a division, at an individual store, or in a particular geography.

A company needs a consistent vision that all team members execute. And corporate visions can change.

But be careful about lurching—be careful about making sweeping changes to execute a new strategy. (In times of severe economic challenge, of course, expect some lurching, some fast change. And respond to, and lead, that change. But think of that as more of an exception in your career, rather than something you should execute frequently.) Find where a new corporate vision and consistent execution find common ground. Where does the new corporate vision naturally create change in an area of the company? That's probably a good place to start. Where does the new strategy build on areas of current executional or market success? That's another good place to start. Don't just change areas of weakness—build your areas of strength.

The Advancement of Change Will Be Uneven

Most of us are trained to think linearly about change and progress. As you improve execution, change will be uneven:

1. **The level of execution.** Since major changes are executed through people, the advancement of execution will be uneven. Training programs, new benchmarking, or a new technology may take time for people to gain a level of comfort with. Those investments (of time or money) may take time to sink in. Some people will adopt the new ways before others will. People who are early adopters will likely buy into the change and have a tendency to execute well.
2. **Lagging perception.** A lot of change we will create will be so someone notices a difference—maybe customers, maybe our own people, maybe distributors. None of these groups will wake up each morning to see if you have changed something. It will take time, it will take some experience,

before they notice a change. In my experience, in consumer goods, customer perceptions can frequently lag months behind a change in reality. Unfortunately, customers will notice changes for the worse much quicker than they notice changes for the better.

3. **Initial failures.** Frequently, when trying to build significant new executional strength, the first challenge is figuring out what to change. Where is the failure point that is causing the execution to lag expectation? And what will be effective at improving the situation? Since money is always limited, finding a cost-effective way to improve execution is always part of the equation. That means that frequently (1) the first thing you fix might not "move the needle," and (2) you may fix a couple of things, and benchmark each one of them, to figure out which fix moves the needle.

As we build execution, we will benchmark our success. Generally, people will use some sort of a "run" chart—the basic chart that has one series of numbers up the side and another series of numbers (or time) along the bottom. A very solid, linear chart.

Expect, as you benchmark against new execution, to see some limited successes before that level of execution settles in as the new standard. Essentially, that is what you are working to accomplish—a new executional standard. Celebrate early wins, but don't become unsettled if you occasionally give back some of those early successes.

In Summary

Figure out what you are willing to give up to be successful.
Practice and prepare.
Persevere through the uneven advancement of building better execution.

Bibliography

Books

Burlingham, Bo, *Small Giants, Companies That Choose to Be Great Instead of Big*

Fair, Charles, *From the Jaws of Victory*

Macksey, Kenneth, *Military Errors of World War Two*

Michaelson, Gerald and Steven, *Sun Tzu for Success* (Adams Media)

Pardoe, James, *How Buffet Does It*

Phillips, Thomas, *Roots of Strategy*

Underhill, Paco, *Why We Buy: The Science of Shopping*

U.S. Army's Infantry Journal's, *Infantry in Battle*

Newspapers and Magazines

Fast Company, September 1997

Forbes, January 9, 2006

Human Resource magazine, Fall 2005

Inc. magazine, July 2004

Multi Channel Merchant, February, 2006

New York Times, July 17 and December 18, 2005

New York Times Magazine, December 25, 2005

Retail Week magazine, March 2005

Wall Street Journal, January 21–22, 2006; November 2005

Internet

www.about.com

CNN Online/Money, 12-23-05 (Netflix quotes)

www.coors.com

www.ordham.edu

www.fortune.com

www.ft.com (Cisco quote)

www.intel.com

www.netflix.com (Netflix financial releases)

www.oup.co.uk (Oxford University Press)

www.wikipedia.com

www.woopidoo.com (Iacocca and Oprah quotes)

Index

R.C.L.
MAI 2008
A